Insect Appreciation

Third Edition

F. Tom Turpin
Purdue University

THE ENTOMOLOGICAL FOUNDATION

KENDALL/HUNT PUBLISHING COMPANY
4050 Westmark Drive Dubuque, Iowa 52002

Chairman and Chief Executive Officer Mark C. Falb
Vice President, Director of National Book Program Alfred C. Grisanti
Editorial Development Supervisor Georgia Botsford
Developmental Editor Tina Bower
Prepress Project Coordinator Sheri Hosek
Prepress Editor Charmayne McMurray
Design Manager Jodi Splinter
Designer Deb Howes
Senior Vice President, College Division Thomas W. Gantz
Vice President and National Field Manager Brian Johnson
Managing Editor, College Field Paul Carty
Associate Editor, College Field Keith Kropp

ISBN 0-7872-9463-2

Printed in the United States of America
10 9 8 7 6 5 4 3 2 1

Thanks to . . .

Don C. Peters who provided me the opportunity to study insects in graduate school, and to

Pat Randolph and Michael Meyer who have served as invaluable teaching assistants in entomology at Purdue University and who have assisted in the preparation of this book.

Contents

Foreword

The goal of this book is to help students develop an appreciation of insects and an understanding of the impact of insects on humankind and the biosphere.

Everyone knows an insect when they see one. Insects are everywhere and human-insect encounters are common. Frequently such encounters are of the negative kind: insects that bite or sting; insects that destroy crops, lawns, or flowers; insects that infest our homes; or insects crawling in the salad at a picnic. The percentage of insects that cause these negative opinions is very low, but the small percentage that are pests tempers our attitude to insects in general. The prevailing human attitude toward "bugs" is one of disgust.

Over the years, courses directed to the study of insects have generally been designed to teach the in-depth biology of insects or the techniques to control pest species. As a rule such courses have limited enrollments because the material is "more than students really want to know about the subject."

The fact remains that, whether we like insects or hate them, almost everyone is interested in learning more about these fascinating creatures.

Because insects are a pervasive part of our daily lives, insects have provided inspiration to many people in the arts. This tie to the humanities, coupled with the widespread, although admittedly reluctant, interest of people is the inspiration for this book. This book provides information to reluctant learners about insects and the importance of insects to our daily lives.

Including a healthy dose of humanities in an attempt to teach scientific concepts is not new. I was influenced by the eminent ant specialist William Morton Wheeler who was a strong proponent of coupling the arts and science in teaching. Wheeler discussed his philosophy of teaching biology in his address as president of the American Society of Naturalists (December 29, 1922) entitled "the Dry-rot of our Academic Biology." In his address he stated:

> *". . . surely we should realize, like the amateur, that the organic world is also an inexhaustible source of spiritual and aesthetic delight. And especially in the college we are unfaithful to our trust if we allow biology to become a colorless, aridly scientific discipline, devoid of living contact with the humanities."*

The broad concepts associated with the study of insects are introduced in a good-humored and understandable manner through short newspaper and magazine style columns. I hope the students who read this book will learn some interesting things about insects—if for no other reason than to become a better educated person!

About the Author

A native of Kansas, Dr. F. Tom Turpin received his undergraduate degree from Washburn University where he majored in biology and played basketball and baseball. He received his Ph.D. in entomology from Iowa State University and has taught entomology at Purdue University since 1971. He has a partial appointment with the Center for Instructional Excellence at Purdue.

Dr. Turpin currently teaches one of the most popular courses at Purdue University. This introductory course on insects called "Insect Friend or Foe" draws over 450 students each semester. He has taught a graduate level course in pest management, bee keeping, honey bee biology, the history of agriculture and currently teaches a course on insects in prose and poetry.

The "Bug Bowl" at Purdue was created by Dr. Turpin as a celebration of insect science. The event has drawn over 20,000 visitors each year and attracts media attention from around the world. He has been featured in numerous news stories including articles in *Time* and *People* magazines. He has appeared on TV and radio programs including the "Guiness World Record Show," "Good Morning America," and "Prairie Home Companion." Tom writes a bi-monthly column "On Six Legs" which appears in over 40 newspapers and does a monthly radio call-in program on WBAA public radio.

In addition to numerous teaching awards, Dr. Turpin was President of the Entomological Society of America (1992), received the Certificate of Distinction from the Purdue Agricultural Alumni Association (1991) and was honored as an Alumni Fellow at Washburn University (1993).

UNIT 1

Entomology and Insects in Perspective

CHAPTER 1

The Study of Insects

Entomology

Entomology is the study of insects. The word entomology is based on the Greek *entomon* which was used to describe the creatures we know today as insects. The Latin word *insectum* which is based on *insecare*—meaning to cut in, a reference to the cut-in appearance of these animals—is the basis for the English word **insect**. Another term sometimes used for insects is **hexapod** which is based on the Greek *hex* (six) + *podos* (foot)—a description of the six-legged nature of these creatures. The study of insects could just as well be called hexapodology or insectology but entomology is the word that has come to be used to describe this branch of science.

Why is there a science called entomology? The study of insects could be a part of biology. To be sure a biology course would be incomplete without including insects. However, the specific study of insects as the science of entomology can be justified in several ways including the following:

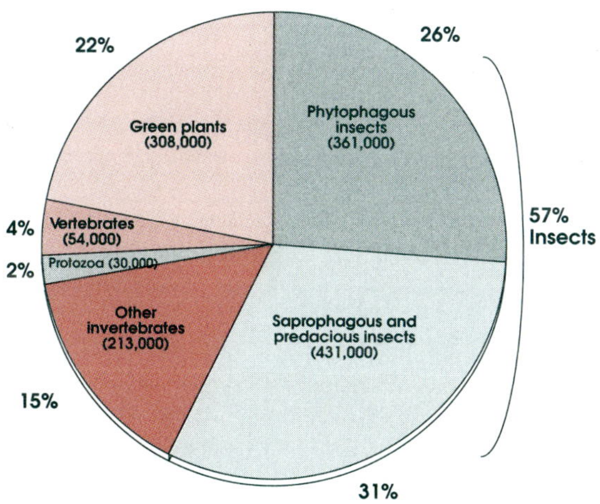

FIGURE 1.1. *The Biological Pie*: Insects, other animals, and plants.

Insects are the dominant group of animals on the surface of the earth and in fresh water (Figure 1.1). There are over 1 million kinds (species) of insects described and many more (some say as many as 10 million) that have not yet been named. Insects make up about 80% of all known animals. It has been estimated that there are as many as 230 million insects on the surface and in the top 9 inches of soil in an acre of meadowland. They play essential roles in the balance of nature including that of predators, parasites, scavengers, and as food for other animals (and plants).

Insects represent a biological system vastly different from ourselves. Their physiology and structure are completely different than ours. They fly and are cold blooded. They don't think and they survive in conditions that humans would find totally intolerable. The "alien" insects and humans in many ways represent the pinnacles of evolution on the earth.

Many insects are valuable to us. Insects produce goods such as honey and services such as control of undesirable pests. They also are the pollinators of many of the fruits and vegetables that we eat. Insects are research animals used in scientific inquiry in many areas including the study of medicine.

Some insects are harmful. While less than 5% of the insect species are considered harmful, those that are cause considerable loss each year. Insects destroy our crops, animals and our possessions, they transmit human diseases, and just plain annoy us with their presence.

Insects are interesting creatures. Because of their diversity of form and living habits many people (other than students of entomology) are fascinated by them. Insects add to the natural beauty around us.

Entomologists—People Who Study Insects

Entomologists are people who concentrate on the study of insects either as a career or as a hobby. Most people who practice entomology as a hobby are interested in insects because of the beauty and diversity of these creatures. They are primarily collectors and taxonomists (classification, especially of plants and animals according to their natural relationships) of insects.

Most entomologists who make a living working with insects are employed by colleges or universities, by federal governments, or by companies dealing with pest control or products for pest control (the pesticide industry). Professional entomologists also work for museums and zoos, industries other than pest control, and as consultants working to solve insect-related problems for clients.

An understanding of insects is important to anyone who is an agriculturalist, naturalist, or biology teacher. There are also many scientists who do biological research at either the cellular level or organismal level who use insects as their model organism.

A Short History of Entomology

Early study of insects was associated with need and the earliest records of entomological activity are from East Asia. In 4700 BC silkworms were being grown for their silk, which was used for guitar strings and fishing lines. In ancient China honey bees were maintained for their honey and wax. Ants were also employed by the Chinese people as biological control agents for insect pests on citrus.

In the Middle East early entomology was associated with magical powers that the people associated with insects. King Menes in the first dynasty in Egypt made the oriental hornet the symbol of his kingdom. In Biblical times the Philistine city of Ekron had a god named Ba'al Zebub—the god of the flies. The sacred scarabaeus is commonly associated with the ancient Egyptian civilization. This scarab beetle was the personification of their sun god.

During the Middle Ages, a time when science was rejected and symbolism flourished, insects were not forgotten. The fly became the symbol of the devil and sorrows, the ant was the symbol of

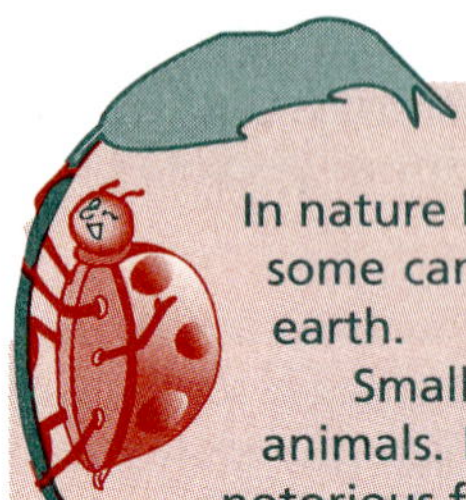

Small Is Beautiful to Bugs

In nature bigger isn't always better, especially when it comes to insects. Even though they are small and some can't be seen without a magnifying glass, insects are among the most successful creatures on earth.

Smallness has its advantages. For instance, insects can hide in cracks and crevices too small for most animals. Many insects escape death simply because they are overlooked due to their size. Gardeners, notorious for their desire to smash insect pests, frequently don't even know insects are around until damage begins to appear.

Another advantage to being small is that the smaller you are the less food you require. Many insects live their lives on food wasted by larger animals. When it comes to food and insects, a little dab will do 'em.

A little wasp, which is a parasite of eggs of other insects, is only about 0.3 mm long. On the other hand, one of the largest insects is the Goliath Beetle. This mammoth among insects weighs about 1.5 ounces. A weight that makes it a little heavier than hummingbirds and shrews, the smallest birds and mammals.

Their small bodies also make it possible for insects to do some marvelous things. For instance some grasshoppers have been observed to jump nearly 10 times their height and 20 times their length. If humans could do as well, just think what the world records for the high jump and long jump might be! Some beetles can drag 120 times their body weight. An average human by comparison would be able to drag about 9 tons.

Although it's an advantage for insects to be small, the way they're built means they can't be much larger. For instance, insects have an exoskeleton which would become too heavy for large animals to carry. Also, insects supply oxygen to their cells though a series of tubes that carry air to all body tissues. Larger animals cannot use this approach. They depend on blood in a circulatory system to supply the oxygen needed by tissues.

To insects being small is indeed beautiful. It is a key to the success of these creatures. How successful are they? In the United States it has been estimated that insects average about 400 pounds per acre compared to the weight of humanity of only 14 pounds per acre. They may be small, but what they lack in size, they make up in numbers.

the provident worker, the bee stood for virginity and wisdom and the moth represented the temptations of the flesh.

The 1700s was a time when scientists were naturalists and they began to systematize and describe things in nature including insects. It was during this time that Carl von Linne, known today as Linnaeus (Figure 1.2), founded the science of systematics. Linnaeus started the classification of insects and he named thousands of insects. Many of the names given to insects by Linnaeus remain today and such names are indicated by an "L" following the name. Another entomologist who described and named many species of insects was J. C. Fabricius, a student of Linnaeus.

FIGURE 1.2. Carolus Linnaeus.

Early American entomologists were biologists or practiced entomology as an avocation, a hobby. Some entomologists consider Frederick Valentine Melsheimer (1749–1814) to be the "Father of American Entomology." Melsheimer was an ordained minister and came to America as the chaplain in a regiment of Hessian dragoons. He remained in America as a minister but practiced entomology throughout his life.

FIGURE 1.3. Thomas Say.

Another early American entomologist was Thomas Say (1778–1834) (Figure 1.3). He was from a wealthy Philadelphia family in the drug business. He helped found the Philadelphia Academy of Science and eventually moved to Indiana as a participant in the socialist community at New Harmony. He died at New Harmony and is buried there.

Charles Valentine Riley (1843–1895), known as C. V., was one of the most controversial figures in the history of American entomology. It was through his efforts that entomology became part of the U.S. Department of Agriculture. He was a pioneer in the practice of biological control and is known for his efforts to introduce the Vedalia beetle, a ladybird beetle, into the U.S. for control of the cottony cushion scale on citrus.

One of the first teachers of entomology was John Henry Comstock (1849–1931). J. H. Comstock spent his entire teaching career at Cornell University.

The Success of Insects

It can be argued that insects and humans are the two most successful creatures on earth. Of course humans represent a single species and "insects" many species but the comparison between the two is frequently made. We are successful primarily because of our ability to think, to reason; and this, combined with our ability to use tools (it's the opposing thumb that makes the task easier) allows us to manage our environment. Insects on the other hand manage to exist in their environment. Indeed they are very successful. The success of insects hinges on several biological considerations including the following:

Skeleton

The insect has an exoskeleton that is very adaptable. It provides protection against the outside world and especially against moisture loss. It was the skeleton that allowed insects to invade the land before most other animals.

Size

Insects are small, at least when compared to many other animals including humans. Some of them are even microscopic. This small size has allowed them to creep into and live in spots inaccessible to other organisms.

Flight

Insects use flight to escape enemies and to inhabit new areas.

Metamorphosis

By changing forms from the immature to the adult stage many insects are able to utilize different food resources.

Reproduction

Insects can produce tremendous numbers in a short time. This is due to two aspects of their reproduction. They lay lots of eggs and have a short generation time.

These biological considerations make insects adaptable organisms. In general it is the adaptability of insects that is the key to their success. They can respond quickly to changes in environmental conditions or even control measures such as pesticides.

In History of Biology, Insects Were There

The history of biology, like nature itself, has lots of insects. For instance, around 330 B.C. the great Greek philosopher and biologist Aristotle put together several books about animals. Aristotle's writings brought together some of the earliest information about what people knew about the nature of life at the time.

These works included material on the habits of fish, reproductive activities of octopi, and biology of whales, porpoises and dolphins. In one volume, called *Historia animalium,* Aristotle included information about honey bees. Most scholars believe that the material was written by someone else, so we think of Aristotle as the editor of the book.

Nonetheless, the material was provided by a bee master, as the individual was called, who knew a lot about bees. The work includes accurate descriptions of a double-sided comb that contains either honey or grubs.

There is a description of drone flights and the ruler leaving with a group of bees. That is what we call "swarming" today. It is also mentioned that bees only visit one type of flower on each nectar-collecting trip and that bees that sting die.

One major mistake is that the writer states that the ruler is a male. A strange conclusion, since the writing indicates that the drone is a male and the workers are female. This gender error about the ruler persisted in literature through the time of Shakespeare—proving once again that some concepts die hard!

By the 16th century, there was a recognized need to organize available scientific knowledge. The individuals of the time who tried to assemble the knowledge were known as encyclopedists. One of the group, Aldrovandi, published an illustrated treatise on insects. Another, the Londoner Thomas Moufet produced his masterpiece, also illustrated, titled "Theatre of Insects."

Marcello Malpighi was a professor of medicine at Bologna University in the 1600s. Malpighi is well-known for his work on the capillary system associated with the circulation of blood. But some of his best work was on an insect. He dissected the silkworm under the microscope and discovered that it had internal structures. He showed that silkworms lack lungs but breathe through a series of holes and tubes. He was correct.

Malpighi's treatise on the silkworm was the first on an invertebrate. In fact, the insect structures that are equivalent in function to the human kidneys are called Malpighian tubules. How's that for a long-lasting scientific honor?

Another historical scientist who dissected insects was Swammerdam. He produced accurate drawings of the internal workings of mayflies and his anatomy of the honey bee is remarkable by almost any standard. He also described the development of dragonflies and mosquitoes

Many people recognize that Leeuwenhoek was a classical microscopist. It is said that he wandered through nature with his microscope, finding new wonders at every turn. Included was his work on the compound eyes of insects. He also discovered that aphids did not have eggs in their bodies but young that looked like the adult.

Leeuwenhoek also did work on the development of the ant and the flea. One of his discoveries was that of a mite parasite on the larvae of fleas. This discovery led to the lines of the satirist Jonathan Swift, which is widely quoted:

So naturalists observe, a flea
Has smaller fleas that on him prey;
And these have smaller still to bite 'em;
And so proceed ad infinitum.

It is not surprising that early scientists discovered insects. Because of the numbers of insects, they are hard to miss even if they are small.

Many People Are Entomologists at Heart

Entomologists, by definition, study insects. That includes all kinds of people.

Researchers who use insects as model animals are entomologists. So are pest control operators who get rid of unwanted insects. Hobbyists who make insect collections, and gardeners who grow plants to attract butterflies certainly are interested in insects.

Teachers who include insects in their lesson plans are studying insects. Parents trying to figure out the name of a caterpillar captured by their young son or daughter, willingly or not, are doing some serious insect study. All are entomologists!

Actually, at some time or another, nearly everyone devotes a little time to studying insects. Some are well-known people who are said to have been students of insect behavior.

For instance, the great Greek philosopher Aristotle studied insect behavior. His observations were accurate enough to allow him to conclude that the "sting" of flies is in the mouth while that of the bee is associated with the rear.

Some of the wisdom of King Solomon is associated with his knowledge of insects. His observations of ants led him to conclude that they might provide a role model for humans. King Solomon said, "Go to the ant thou sluggard, consider her ways and be wise!"

Many of those who today are considered pioneer entomologists studied insects as an avocation—a hobby—rather than as a job. Historically, early entomologists were either preachers or medical doctors. These were two professions that were available at the time to people with a classical education.

In his book *American Entomologists,* Arnold Mallis includes the more famous of these early entomologists. One was the Rev. Frederick Valentine Melsheimer. Melsheimer was a Lutheran clergyman who made the first important collection of insects in the United States. He also wrote the first important entomological work in the nation: *A Catalogue of Insects of Pennsylvania.*

Mallis concludes relative to Melsheimer, "As can be imagined, the Reverend's activities in pursuit of bug, beetle, or butterfly were a source of amusement to his parishioners, who considered it a harmless eccentricity on the part of their pastor."

Melsheimer's son, a medical doctor, continued his father's interest in insects and maintained the Melsheimer collection of insects until it was sold to Harvard University.

Many physicians have become well-known entomologists. One such individual was William LaBaron. LeBaron served as state entomologist of Illinois, but, prior to that time, he was a physician like his father and grandfather. He became "known far and near for his skill in diagnosis of disease and surgical operations."

Today, entomology is a profession where people make a living by studying insect biology. However, many people still engage in insect study as a hobby. That is why most entomological organizations include a large number of insect hobbyists.

It is still true today, as it has been throughout the annals of history, that people pay attention to insects. After all, insects outnumber us several thousand to one, so it makes sense to know something about them!

Bugs Bad Rap

"Bugs bad rap" could be a music critic's negative review of the latest album of a rap group. Or it could be a Warner Bros. release of a new Bugs Bunny cartoon. Bugs bad rap, however, is a commentary on the negative attitude that most people have toward insects.

In her Pulitzer prize winning book, *Pilgrim at Tinker Creek*, Annie Dillard summarizes that attitude with these lines: "Fish gotta swim and birds gotta fly; insects it seems, gotta do one horrible thing after another. I never ask why of a vulture or shark, but I ask why of almost every insect I see. More than one insect—the possibility of fertile reproduction—is an assault on all human value, all hope of a reasonable god."

Annie Dillard has hit the nail on the head. Most of us view insects in the light of the bad things they do. To be sure, insects bite and sting, they transmit diseases to plants and animals, they destroy our crops and possessions, and they generally make nuisances of themselves.

The insect world, in the view of humans at least, really lends credence to the old saying that "One bad apple can spoil a barrel of good ones!" In fact, less than five percent of all known insects are bad. They are the ones that grab the headlines and spoil our patio barbecues.

As Paul Harvey might say, the other 95 percent of insects are "the rest of the story." Indeed insects provide valuable goods and services to humankind. The most obvious are products like honey and silk. But insects provide animals for research in medicine, genetics, population dynamics and basic biology.

In nature, insects are important to the grand scheme of things. They provide food for many animals. They pollinate the plants that provide many of our fruits and vegetables. They take a leading role in recycling. Insects are johnny-on-the spot for road kills. Otherwise roadsides would be littered with dead animals. Termites turn dead wood into humus. And the billions of tiny insects that inhabit the soil help keep it aerated and "alive."

In the uneasy relationship between humans and insects, the insects provide far more benefits than harm. Consequently, "bugs" do get a bad rap.

Entomology and Entomologists

1. About what percentage of all named organisms are insects?
2. What are five good reasons to study entomology separate from biology?
3. What is the etymological (the origin of words) basis for the word entomology? Insects? Hexapod?
4. What are the common names of three insects used extensively by the Chinese people nearly 5,000 years before the time of Christ?
5. Name an early American entomologist who was associated with the socialist experiment at New Harmony, Indiana.
6. What are five biological factors that contribute to the success of insects?

CHAPTER 2

Insects and Other Animals

Insects have been represented on the earth for over 350 million years. Among the earliest of the insects were the cockroaches. Cockroaches were around when the first amphibians roamed the earth. These roaches survived the glaciers that wiped out the dinosaurs. By the time birds where taking flight for the first time insects were flying everywhere. Seed plants in part were able to develop because insects were there to act as pollinators. When mammals showed up insects—the fleas—developed to feed upon the hairy creatures. Lice are one of the more recent developments in the insect world. It is said that lice owe their lifestyle to the development of civilization, or the lack thereof. When we live in crowded conditions we provide the kind of habitat in which lice thrive.

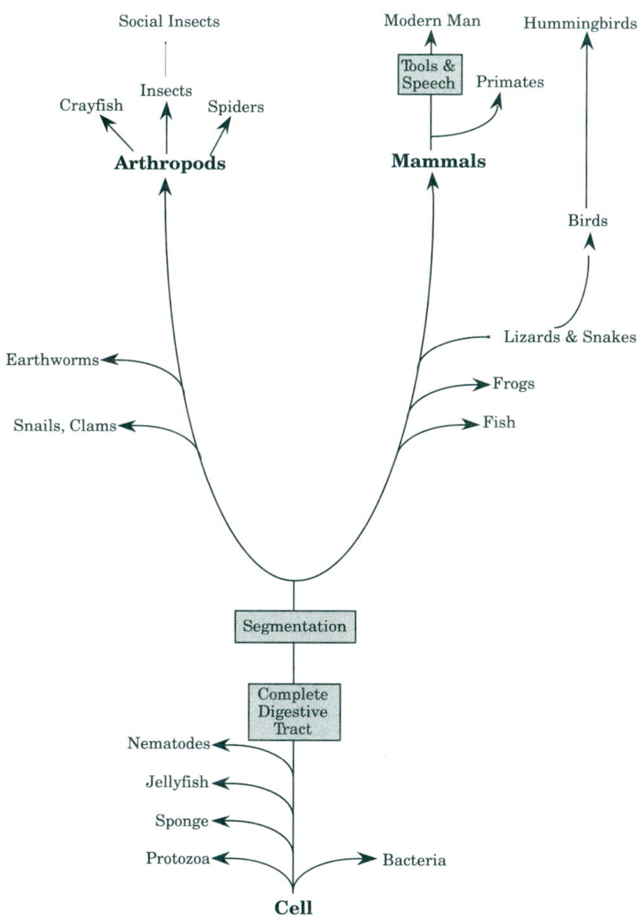

FIGURE 2.1. *Evolutionary Relationships among Animals.*

Insects have been around for 350 million years. Humans for only 10,000 years! Insects certainly were on the earth long before humans and will probably be here long after we're gone. Such was the sentiment of the entomologist W. J. Holland, who concluded his book "The Moth Book" with the following which he entitled "The End."

> *When the moon shall have faded out from the sky, and sun shall shine at noonday a dull cherry-red, and the seas shall be frozen over, and the ice-cap shall have crept downward to the equator from either pole, and no keels shall cut the waters, nor wheels turn in mills, when all cities shall have long been dead and crumbled into dust, and all life shall be on the very last verge of extinction on this globe; then, on a bit of lichen, growing on the bald rocks beside the eternal snows of Panama, shall be seated a tiny insect, preening its antennae in the glow of the worn-out sun, representing the sole survival of animal life on this our earth,— a melancholy "bug."*

Certainly the success of insects as a life form is secure. In fact it has been suggested that the pinnacles of biological evolution to this date are three: the human being with our ability to think (although this is questionable in many instances) coupled with our ability to use tools; the energetics of the hummingbird; and social insects. Even "archy" the cockroach recognized the latter accomplishment when he wrote:

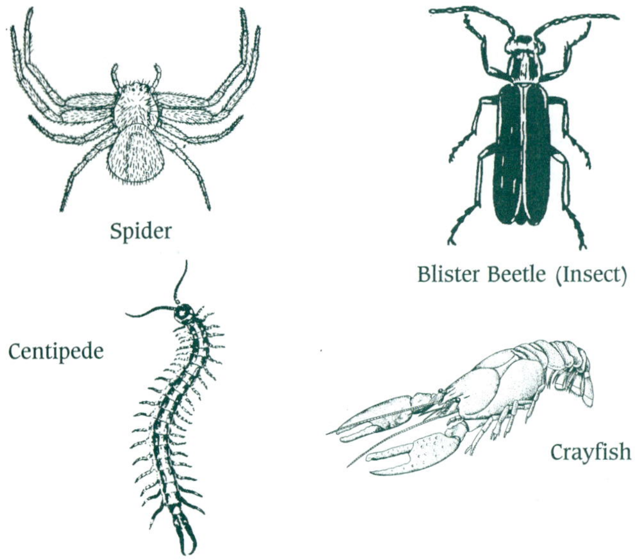

FIGURE 2.2. *Common Arthropods.*

the bees got their
governmental system settled
millions of years ago
but the human race is still
groping

ARCHY

From archy and mehitabel by Don Marquis 1927, Doubleday & Co.

Insects are part of the phylum of animals called *arthropods*. Arthropods are named for one of their dominant physical characteristics—that of jointed limbs (from the Greek *arthr* = joint; *podi* = foot). Arthropods are also covered with a hard shell or skeleton known as an exoskeleton. In addition to insects, arthropods include such common creatures as spiders, ticks, millipedes, centipedes, lobsters, crayfish and crabs. Crustacea (lobsters, crayfish, shrimp) are the dominant arthropods in water, and insects are the dominant arthropods on land.

The major arthropods can be separated according to these characteristics: insects have 3 pairs of legs, 3 body divisions, and usually have wings. Spiders, mites, and ticks have 2 body divisions (they appear to have lost their head), 4 pairs of legs, no antennae, (although sometimes palps on the mouth may resemble them) and no wings. Lobsters and crayfish have several pairs of legs, 2 pairs of antennae and no wings. Centipedes and millipedes have many pairs of legs and no wings.

Insects and Other Animals

1. Name a key geological event that was associated with the development of these insects: butterflies; fleas; lice.
2. To what animal phylum do insects belong? What are the common names of other animals in this phylum?
3. Gossamer is associated with what arthropod?

THE GEOLOGIC RECORD OF INSECTS

Geologic Period	Millions of Years	Key Events	Insects
Cambrian	600	Marine Invertebrates	
Silurian	425	Land Plants	
Carboniferous	310	Insects, Amphibians	Roaches, Mayflies
Permian	280	Glaciers	Dragonflies, Cicadas, Lacewings
Triassic	230	Reptiles	Ants, Bees, Bugs
Jurassic	185	Birds, Dinosaurs	Flies
Cretaceous	135	Seed Plants	Butterflies, Termites
Tertiary	60	Mammals	Fleas
Recent		Civilization	Lice

Compost Piles Are Alive with Arthropods

Any gardener worth their spinach or their creeping phlox possesses a compost pile. The compost pile is not the centerpiece of most gardens, but it does play a very important role. That is where the spent blossoms, pruned branches and pulled weeds are consigned.

But the compost pile does more than get things out of sight. It recycles them. This year's broccoli leaves become nutrients for next year's black-eyed Susans.

Good compost piles are more than a heap of dead plants. They are alive with the sounds of chewing. At least, they would be if we could hear the sounds of tiny mouthparts chowing down on discarded plant tissue. You see, compost piles are filled with tiny creatures that make their ecological living by feeding on dead stuff.

Most of these inhabitants of the compost pile are arthropods, creatures with exoskeletons and jointed legs. To be sure, earthworms are also common in compost and play an important role in breaking down the plant tissue. But by sheer numbers the arthropods dominate the compost heap.

Any sharp-eyed gardener turning over the compost will notice these creatures that shun the light. First there are literally millions of little white or gray things that seem to pop from place to place. These are tiny, wingless insects called Collembola.

Collembola are sometimes called springtails. That name is based on a tail-like spring device that they use to catapult themselves from place to place. Collembola are the most common insects on the Earth, but because of their size and reclusive habits they go mostly unnoticed.

Sowbugs also are common in the heap. They are sometimes called roly polies or pillbugs. Sowbugs aren't insects, although most people regard them as such. They are crustaceans, as are crayfish, lobsters and crabs. Most crustaceans live in the water, but sowbugs can be found on the land doing the same thing as their water cousins—eating dead stuff.

Millipedes also find the compost pile to be a wonderful home. With two pairs of legs on each body segment, they have a lot of appendages. Not quite a thousand, as the name millipede suggests, however. Millipedes are also recyclers, feeding on decaying vegetable matter.

Everything that shows up in compost isn't feeding on dead material. Take that many-legged cousin of the millipede, the centipede. It has one pair of legs on each segment and moves very fast. Its speed suggests it doesn't feed on decaying matter. In fact, it is a predator and makes its living feeding on other arthropods.

Black beetles can sometimes be seen running from the light when the compost is turned over. These beetles are known as carabid or ground beetles. They sometimes feed on decaying plant material, but some are predators, preferring to chow down on other arthropods.

This is also the case with the spiders that hang around the compost pile. The spiders, including the granddaddy longlegs, are happy to make a meal out of some of the creatures that live there.

The compost pile might be where gardeners consign their dead plants, but it is not a dead place. Arthropods by the millions are busy working day and night turning that plant tissue into compost. It's one of nature's dirty little jobs, but someone or some arthropod has got to do it.

These Arthropods Have Too Many Legs to Be Insects

Some people call them insects or worms. But these many-legged creatures are not insects or worms. Millipedes and centipedes are, however, arthropod relatives of insects.

Millipedes and centipedes are similar in structure. They are worm-like, with legs on each segment. Together, millipedes and centipedes are sometimes known as myriapods—a term that means many feet. Myriapods have nine or more pairs of legs.

While the name millipede suggests that they have 1,000 legs, it isn't true. But millipedes do have lots of legs, more than 30 pairs. The number of legs depends on the number of body segments. Millipedes have four legs on each segment.

A common species of millipede found in the Midwest has 50 segments and 200 feet. There is a giant tropical millipede that has up to 320 feet arrayed on 80 segments.

All of those legs require some coordination. So millipedes keep from getting their feet tangled up by moving them in an organized fashion. When a millipede is walking, its legs move in a peristaltic wave motion. This works when a small number of legs on both sides of the millipede's body move forward together. Once these legs are forward the next group of legs moves and so on, until the wave reaches the front of the creature.

This approach to walking is visually similar to the wave that fans at athletic contests sometimes produce. In both cases, the participants—fans at athletic events and legs of millipedes—wait their turn.

Peristaltic wave motion is also found in the digestive tract of mammals. The muscles of the digestive system contract in rhythmic sequence to move the food mass through the body.

Millipedes are slow-moving creatures. They are not in a hurry to get from place to place. Since they feed on decaying vegetable matter, it is not important to move fast to find a meal. Their food waits on them to show up.

Millipedes also have a bad taste. So they generally don't have to run away to avoid becoming a meal for other animals. But if threatened, millipedes just curl into a ball, hiding their head and legs. This action makes it difficult for predators to find a place to get started eating the millipede.

Centipedes look and behave differently than millipedes. Centipedes have only one pair of legs on each body segment. They also move quite rapidly. Speed is important to centipedes, since they feed on other arthropods and have to catch their prey.

In addition to their speed, centipedes also possess a poison, which is used to kill prey. The scientific name for the centipedes, Chilopoda, literally means poison feet. This is based on the fact that centipedes bite with a pair of modified legs.

The name centipede literally means hundred-legged. But, like the millipedes, the name exaggerates the actual number of legs. Most centipedes have 20 to 30 legs.

So when faced with a many-segmented, many-legged worm, what is it? Well, if it has two long legs on each segment and runs fast it is a centipede. If it has four legs on each segment, crawls slowly and curls up when disturbed, it is a millipede. Good to know, since centipedes can bite when handled!

Traveling on Gossamer without Wings

Spiders are wingless. But the lack of wings doesn't keep these arthropods from traveling. Sometimes spiders walk from place to place. Sometimes they hitch a ride. That explains how a cockroach-eating spider native to South America sometimes shows up in the United States. It travels compliments of imported grapes!

Spiders sometimes fly. Not under their own power, of course, but by a process known as ballooning. Ballooning spiders hitch a ride on their silk as the breeze carries it.

Spider silk floating on wind currents is known as gossamer. It has attracted attention for centuries. The word "gossamer" is an old English term, apparently based on a period of warm weather in November, known as goose summer. That was the time of year when geese were eaten. Late fall is also the time when ballooning spiders are most likely to be seen floating on the breeze.

Airborne spiders are young spiders sometimes called spiderlings. Like all spiders, spiderlings produce strands of silk. These young spiders do not build webs. The newly produced strands of silk wave in the fall breeze like flags on a flagpole. At some point, the young spider lets go. Like a wind surfer being towed by a speedboat, the young spider then becomes airborne.

The process allows the spiderlings to be carried to new locations. Sometimes the landing sites might not be hospitable to spiders. For instance, ballooning spiders have been found far out at sea. Probably not a good destination for terrestrial organisms! Nonetheless, some of the ballooning spiders land in good spots, at least to spiders, and are able to develop into adults during the next season.

Walt Whitman summarizes the process used by the ballooning spiders in these lines of poetry:

"A noiseless, patient spider,
I mark'd where on a little promontory it stood isolated,
Mark'd how to explore the vacant vast surrounding,
It launch'd forth filament, filament, filament out of itself:
Ever unreeling them, ever tirelessly speeding them.
And you O my soul where you stand,
Surrounded, detached, in measureless oceans of space.
Ceaselessly, musing, venturing, throwing, seeking the spheres to connect them.
Till the bridge you will need be form'd, till the ductile anchor hold;
Till the gossamer thread you fling catch somewhere, O my soul."

In her sonnet LXII appropriately named "The Gossamer," Charlotte Smith, nearly 200 years ago, wrote in the first four lines:

"O'er faded heath-flowers spun, or thorny furze,
The filmy Gossamer is lightly spread;
Waving in every sighing air that stirs,
As Fairy fingers had entwined the thread."

While spider ballooning occurs every year, in some years it is quite noticeable. It always occurs on warm, still days late in the fall. To see strands of spider silk floating over an open meadow is an interesting sight. Such was the case in late October when many people in the state of Indiana saw large amounts of gossamer floating about. The natural question was "What is going on?"

A good answer to that question is provided by Anna Botsford Comstock in her book, "A Handbook of Nature Study." She states: "Thus we see that the spiders have the same way of distributing their species over the globe as have the thistles and dandelions."

Dog Ticks Find Humans to Be Suitable Hosts

They are tenacious blood-suckers that are protected with a heavy chitin covering. These creatures can withstand long periods of starvation and can regenerate lost limbs and mutilated mouthparts. They can feed on a wide range of animals, but almost nothing feeds on them.

Ticks! That scourge of the good old summertime. This is not the latest sci-fi thriller from the movie moguls of Hollywood, but one of nature's creatures.

Ticks are not insects. These animals, like insects, are classified as arthropods but are more closely related to scorpions, spiders and mites. Unlike insects, ticks do not have a head and, in the adult stage, have eight legs.

Like insects, ticks have four life stages, and they molt, or shed, their exoskeleton as they grow. The stages are egg, the six-legged larva, which is sometimes called a seed tick, the eight-legged nymph and the adult.

It is the blood-sucking habit of ticks that makes them a problem for humans. Ticks must have blood to develop. During feeding, they attach to their host. Following the blood meal, the tick drops off the host.

So how does a tick go about its life anyway? The newly hatched seed ticks climb upon grass or small trees to await the arrival of a host animal. The young tick will climb on the host and may wander around for a time before finding a place to attach. Once attached, the larvae feed and become filled with blood. When blood-filled, called "engorged," most ticks drop from the host to develop to the next stage.

Then, the tick molts into a nymph and must find another host for a second meal. Play it again, Sam! The nymph feeds and drops from the host to become an adult. Then, the adult finds another host and, after mating and a blood meal, the female drops from the host to lay eggs.

Now the good part, to the tick, that is. When a tick gets on a host and finds a suitable feeding site, it takes a firm grasp with its forelegs. The tick then drives its mouthparts through the host's skin. While feeding, the tick typically voids excretory products, which might contain pathogenic organisms that can enter the host's body through the puncture.

Ticks, as organisms, are well adapted for survival. They lay a large number of eggs—as many as 4,500 for a single female. Ticks are also able to survive for long periods without food or water. One species of tick survived for three years while confined to jars of sand.

The most common tick that humans encounter is the American dog tick. It is usually found in woods and areas of uncut vegetation. This tick will feed on a number of different animals, including dogs, cattle, horses and humans. The deer tick is also of concern since it is the primary vector of Lyme disease. It is much smaller than the dog tick and has no white markings on its back.

The best way to avoid contacting ticks is to stay away from the places where they live. If you are in tick habitats, try not to brush vegetation. Also tuck in your shirt and pull socks over pant cuffs. Application of an insect repellent to shoes and pants also helps.

When returning from outings, it is a good idea to check yourself, children and pets for ticks. Many times you will find the tick before it has found a place to feed. If it has attached, use blunt tweezers and grasp the tick close to the skin; pull with a steady pressure. Then disinfect the bite.

Turn about is fair play. If the tick spends its entire life finding host animals, it makes sense that we can spend a little time finding the tick when it decides to make us a host!

CHAPTER 3

Collecting and Identifying Insects

Insect Collection Equipment

Large insect collections may be made using a small amount of inexpensive equipment. Nets and killing jars are the most essential, although a few other items mentioned later will come in handy and enable the collector to obtain a larger variety of specimens.

The Net

A net serves two purposes. One is to *beat* or *sweep* foliage. This means that the net is swung back and forth so that it scrapes the tops of plants and collects insects that are feeding or resting there. The other use for the net is to collect bees, wasps, butterflies, dragonflies and other flying or wary insects. A suitable net is one that is sturdy enough to sweep plants, yet light and porous enough to be swung through the air. Such a net can either be purchased or made at home.

A net (Figure 3.1) consists of a handle fitted on a heavy wire loop. A bag is fastened to the loop.

FIGURE 3.1. *The Insect Net.*

Select a strong, lightweight handle about 3 feet long. A 0.75 inch dowel rod is ideal for this purpose. Cut two grooves along the sides at one end as shown in Figure 3.2a. These grooves cradle the bent arms of the hoop and should be cut as deep as the thickness of the wire. Make one groove about 3.5 inches long, and the other 2.5 inches. At the end of each, drill a small hole through the handle.

To make the hoop, bend a 4-foot length of about 0.1 inch durable steel wire, preferably piano wire, into a hoop with short arms at each end as shown in Figure 3.2b. Take care that the arms and little hooks at their ends are bent just right to fit along the grooves and into the holes in the handle. After fitting the hoop to the handle and properly attaching the bag, you are ready to make the joint fast between the handle and the hoop.

Wrap the joint tightly with fine wire (Figure 3.2c) or better still fit the handle with a sliding metal ferrule (Figure 3.2d). A short piece of 0.75 inch copper or aluminum tubing will fit snugly over a dowel rod of that diameter.

The bag should be cut out and made cone-shaped as shown in Figure 3.1 and should be twice as long as the diameter of the hoop. This length lets you loop the bag over the rim and thus prevent the escape of insects. An ideal material for making the bag is nylon mosquito netting available from surplus stores. A good quality marquisette or similar material can also be used, but cotton mosquito netting or cheese cloth is not satisfactory.

To construct the net bag, cut out two pieces of net material the size and shape shown in Figure 3.3. Place these together, one directly atop the other and sew from point *a* to point *b* to point *c*. This seam should be about 0.5-inch in from the cut edges. Fold the outer edge (from the seam to the

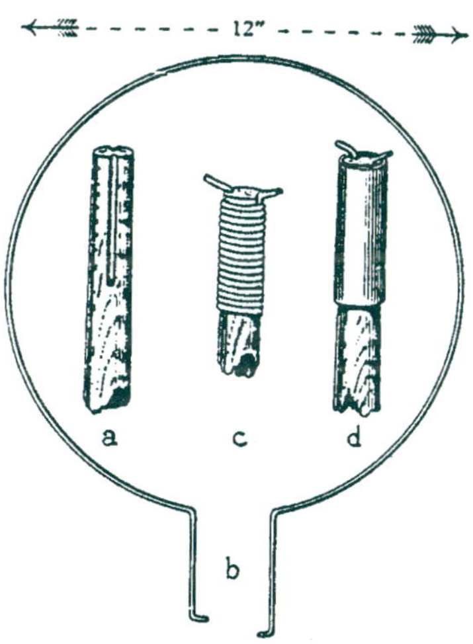

FIGURE 3.2. *The Net Ring.* Two grooves of unequal length are cut on opposite sides of handle at one end (a). These grooves terminate in small holes into which the bent arms of the wire hoop or ring (b) are fitted. The joint may be wrapped with a wire (c) or fitted with a sliding ferrule (d).

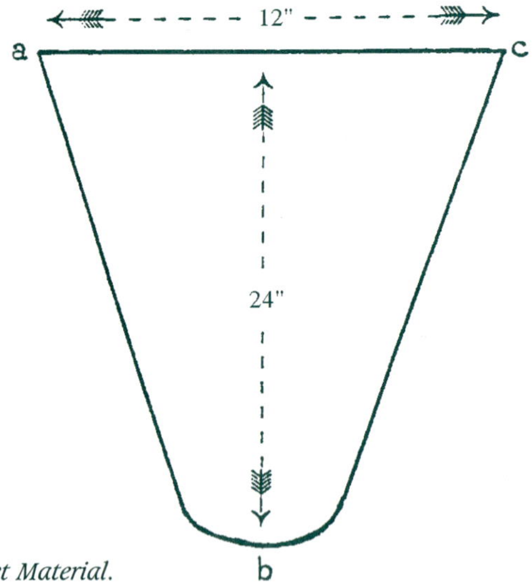

FIGURE 3.3.
Pattern for Net Material.

cut edge) back onto the body of the net and sew around once more, this seam being about 0.1 inch from the cut edges. To make a loop for the wire hoop, use a strip of heavy muslin 5 x 44 inches. Fold it lengthwise to form a hem around the top of the bag. The top edge of the net should be placed between the two sides of the folded muslin. Tuck the cut edges of the muslin in such a way that the edge of the net and the tucked edges of the muslin overlap at least 0.5 inch. Combine these by twice sewing completely around near the middle of the overlap.

The Killing Jars

After insects have been collected, they must be killed quickly without breaking or otherwise damaging them. A killing jar is used for this purpose. Every collector should have at least two such jars—one large enough for butterflies and moths and another for beetles and small insects.

Jars that use ethyl acetate or 1-1-1 trichloroethane as the killing agent can be made at home. The ethyl acetate is safer and excellent for use by young people and beginning collectors. These chemicals can generally be purchased at drug stores. The chemical 1-1-1 trichloroethane is also sold as Engine Cleaning Fluid (use the fireproof one), and may be found in other spot removers. Regardless of the chemical you use, select a jar of heavy glass having a large mouth and a screw cap. Do not use plastic jars.

Ethyl acetate or 1-1-1 trichloroethane jars. Pour about 1 inch of wet plaster of Paris (more for large jars) into the bottom of the jar. Let it harden, and then dry thoroughly in the oven. After removing from the oven, saturate the dry plaster of Paris with the killing agent; pour off any excess liquid that does not soak in. Put the screw cap lid on the jar. Then put a label on the jar saying "INSECT KILLING JAR"—also list the chemical you have used for the killing agent. The jar is now ready for use; however, keep it tightly capped when not in use. When the jar loses its killing strength, dry it out again and recharge.

Cyanide jars. Killing jars containing sodium or potassium cyanide may be preferred by advanced collectors; however, their construction and use is no longer recommended. Collectors requiring cyanide killing jars can purchase them from biological supply houses. This suggestion is offered because restrictions governing the sale of cyanide makes it difficult to obtain.

How to Use Killing Jars

Always keep a piece of clean, crumpled paper in each jar to absorb moisture and to keep the specimens from becoming rubbed or broken. A successful insect collector makes it a practice to mount and label all specimens within a few hours after they are caught. Insects left in the killing jar for a day or two will become soft and ruined, whereas those taken out but not pinned will become too brittle to handle.

Never put damaged or mutilated specimens in the killing jar. Most insects are so plentiful that there is no excuse for wasting time and energy trying to mount and label anything less than a perfect specimen.

How to Pin Insects

Insect pins may be obtained from any dealer in entomological supplies. Do not use common pins, since they will rust and soon ruin what may be valuable specimens. Pins come in several sizes, but number 2s and 3s will be found the most useful.

Any insect that is large enough to support a pin without breaking or otherwise being distorted may be pinned directly through the body. Insert the pin through a definite part of the body from top to bottom (Figure 3.4a). The place of insertion depends upon the type of insect. The following rules have been set up for pinning different types of insects so that the pin may be placed firmly through the heavier parts of the body without destroying important identifying characteristics.

1. Bees, wasps, flies, etc.: Pin through the thorax between bases of forewings and just right of the middle line (Figure 3.4b).
2. True bugs: Pin through the scutellum, which is the triangular area between the bases of the wings (Figure 3.4c).
3. Grasshoppers, crickets, etc.: Pin through the prothorax or "saddle" just to the right of the center line (Figure 3.4d).
4. Beetles: Pin through the fore part of the right wing cover near the center line (Figure 3.4e).
5. Butterflies, moths, dragonflies, etc.: Pin through the center of the thorax between the bases of the forewings (Figures 3.5a and 3.5b).

About 1/4 to 1/3 of the pin should project above the specimen.

As each specimen is pinned, straighten the legs, antenna or wings so they will dry in the desired position. Soft-bodied insects, such as crickets or walking sticks, can be held in position by blocking them up with pieces of light cardboard placed on the pin. These temporary supports should not be removed until the specimen has become thoroughly dry.

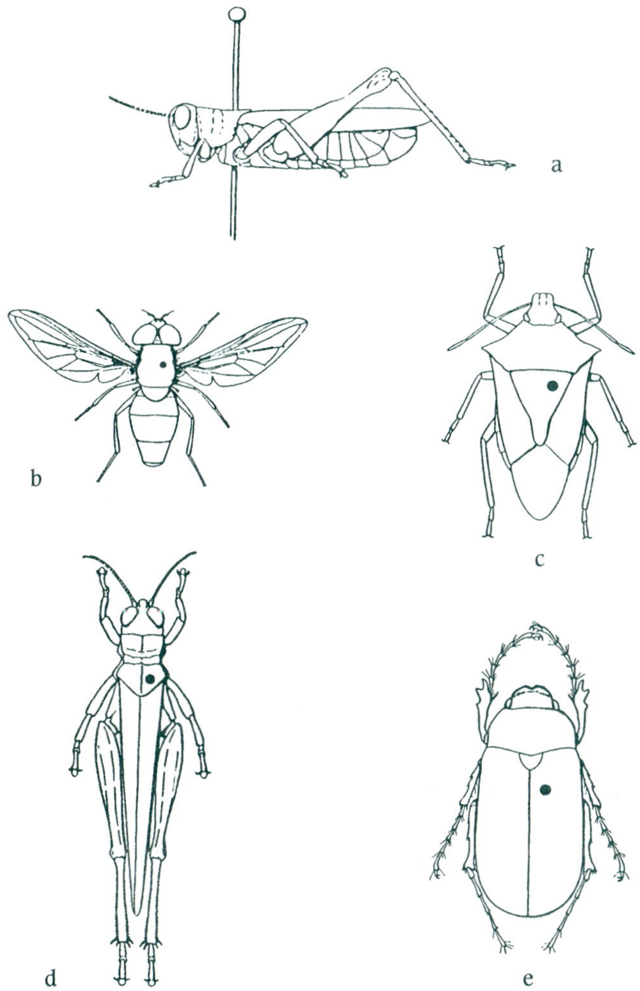

FIGURE 3.4. Side view (a) of a pinned grasshopper. The black dots show the location of the pins in the fly (b), stink bug (c), grasshopper (d), and beetle (e). The pins are inserted through the insects from top to bottom, leaving 1/4 to 1/3 the length of the pins projecting above the backs.

The Pinning Block

The appearance of a collection is vastly improved if all the specimens are placed at a uniform height on the pins. This is easily done by using a "pinning block" (Figure 3.6). Such a block may be made either of wood or metal. Metal blocks, of course, must be purchased, but wooden ones can be made at home.

The wooden block is cut out of a soft piece of wood 1-inch square and 4 inches long, or it may be built up by gluing together four pieces of 0.25 inch finishing lath. Small holes are drilled through each of the four steps, which are 0.25, 0.5, 0.75 and 1 inch high, respectively. After an insect

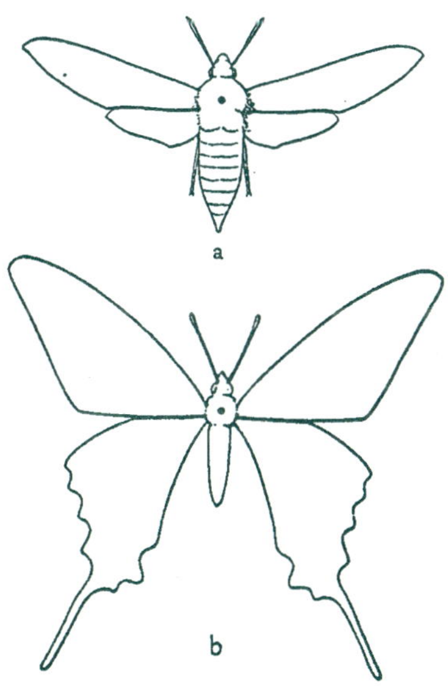

FIGURE 3.5. In both moths (a) and butterflies (b), the pin is thrust through the center of the thorax between the bases of the forewings, just to the right of the mid-line.

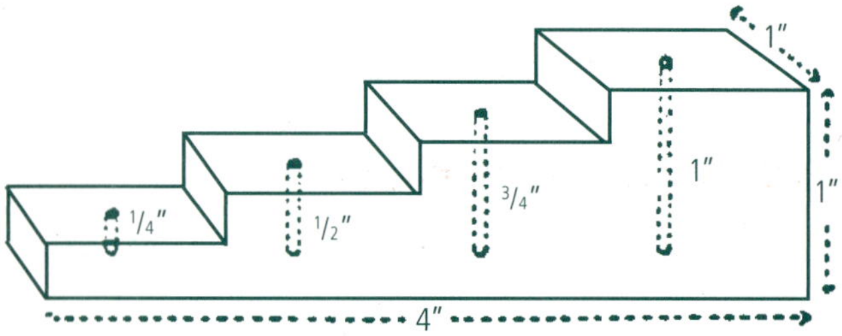

FIGURE 3.6. *The Pinning Block.* This device is used to adjust both insects and labels to their proper heights on the pins.

is placed on a pin, either the head or the point of the pin is inserted in the desired hole and the specimen adjusted to the proper height. The block is also very useful to adjust labels to a uniform height upon the pins.

How to Pin Small Insects

Figure 3.7 illustrates the card point method used to mount very small insects. Cut all points to a uniform size, 0.3 to 0.5 inch long and about 0.1-inch wide at the base. They are easily cut from a strip of filing card paper as wide as the points are to be long. Cut a large supply at one time, and keep them in a box for future use. This will save time and assure that the points will be uniform in size.

Note from Figure 3.7 that the pin is pushed through the base of the triangle and the specimen glued to the point. It is very important that the point be to the left of the pin and the specimen glued so that it will face away from the operator.

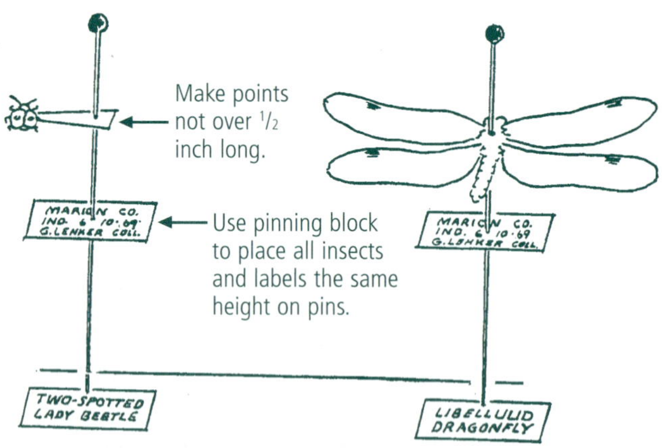

FIGURE 3.7. *Mounting Small Insects and Labeling.* Insects too small to be pinned may be mounted by the card point method (left). All others should be mounted directly upon the pins (right). Note the position of labels on the pins.

Mounting insects on card points is not difficult if you know the correct procedure. First, put the card point on the pin, and place a small amount of clear finger-nail polish remover on the tip. Next, lay the specimen on its back on the edge of a table in such a manner that the pin and point can be turned upside down and pressed lightly against the insect. With a little practice you will learn to lay the specimen so that it will be in the correct position when glued to the point.

The Spreading Board

Butterflies, moths and frequently other insects are mounted with their wings spread. A spreading board is used for this purpose and is an important piece of equipment for the insect collector. Eventually, the collector should have different sized spreading boards to accommodate both large and small insects, but one board will do for the beginner. Adjustable spreading boards are available from entomological supply houses.

The complete spreading board (Figure 3.8) may be made at home, if desired, and is made of the following materials:

1. Two end blocks, 5.5 inches long and 1 inch square.
2. Two soft wood top pieces, 16 inches long, 2.5 inches wide and about 0.5 inch thick. These pieces should be planed down so that they are 0.3 inch thick on one edge.
3. One flat strip of corrugated box paper or cork, 14-inches long, 1 inch wide, and about 0.5 inch thick.

The two top pieces are nailed to the end blocks (Figure 3.8), leaving about 0.5 inch between the thin edges. The cork or corrugated paper is then tacked beneath the top pieces to cover the opening and provide soft material into which insect pins can be inserted.

The slope of the top pieces permits the wings to dry in a slight upward position. This allows for any sag that may occur after the specimen is removed from the board. Spreading boards with flat top pieces will do, but insects must remain upon such boards for a longer time.

The top pieces can also be sloped by carving the top sides of the end blocks into shallow V's.

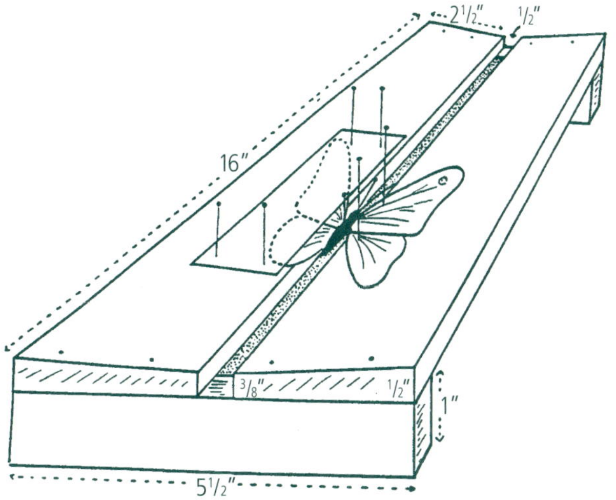

FIGURE 3.8. *The Spreading Board Showing Dimensions.* A pin is inserted through the body into the cork so the specimen rests in the groove. The wings are then pulled forward and held temporarily by pins, as shown on the right, while they are being permanently secured by a strip shown on the left.

An inexpensive spreading board can be made from styrofoam. A groove to accommodate the body of the insect can be carved into a block of this material or pieces of stryofoam can be glued together to make a board similar in size and shape to the wooden one previously described. Boards of this kind do have some disadvantages. For example, they are more likely to rub the scales from the wings as they are being spread.

Spreading the Wings

Before attempting to spread an insect's wings, make sure that the insect is fresh or thoroughly relaxed. Pin the specimen in the usual manner; then thrust the pin through the cork, and push down until the body lies in such a way that the wings are flush with the top pieces. Now, with a fine pin or needle caught behind the heavy front margin of each wing, pull the forewings forward until their hind margins form a straight line at right angles to the long axis of the body. Then hold the forewings in place temporarily by pushing the pins into the wood, and work the hind wings forward in a similar manner until their front margins are concealed beneath the forewings. After you are satisfied with the position of the wings, secure them permanently by pinning strips of paper lengthwise across them.

Most collectors notch these strips of holding paper so the strips will fit around the temporary pins and cover the entire wing surface. After the wings are secured, remove the temporary pins, and set the board aside for 10 days to 2 weeks, after which the specimens may be removed.

How to Label Specimens

To be of any scientific value, every specimen must be accompanied by information including the date and locality of its capture, and the name or initials of the collector. These data are printed on a small label that is placed on the pin beneath the specimen as shown in Figure 3.7. Collections for scientific purposes also have a second label on each pin showing the host or habitat of the specimen. The size of the pin labels should not be over 0.8-inch long and 0.3-inch wide. This is about the smallest size that can be written or printed on with ease. Labels used by entomologists, biologists, and college students are usually much smaller.

Homemade labels should be cut from stiff paper, such as index card stock. It is essential that all labels be the same size; and collectors will find many advantages in cutting a large supply for future use. Those directing the activities of insect-collecting groups might want a printer to set up the locality labels in small type so that only the date and initials of the collector need be filled in by hand.

Always use a pinning block to place labels at a uniform height on the pins.

Display Boxes

Pinned insects cannot be kept in good condition unless placed in boxes to protect them from dust and damage. A standard display box has 18-inch x 24-inch x 2.75-inch outside measurements and has a glass top. Figure 3.9 shows a simple box that is easy to make.

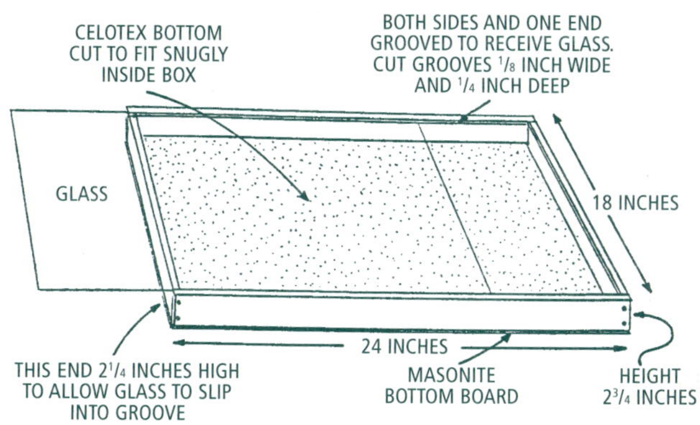

FIGURE 3.9. An inexpensive display box for holding pinned insect specimens.

Flat pieces of celotex or styrofoam provide a handy place for the temporary holding of pinned specimens while they are being labeled, identified, or arranged in display boxes.

Precaution Against Pests

Several types of small dermestid beetles as well as other incidental pests readily feed upon dry insect specimens. To protect against those pests, it is necessary to use an insecticide or repellent in the display boxes. Naphthalene is a good repellent. If collections are infested, paradichlorobenzene (PDB) flakes can be used as a fumigant. Simply scatter a liberal amount of the flakes on the floor of each box, and close the lid.

Insect Collection

1. What are the two most essential pieces of equipment needed for insect collecting?
2. Why is an insect net supposed to be twice as long as it is wide?
3. What are the purposes of a pinning block and a spreading board?
4. What three types of information must accompany a pinned insect if it is to be of any scientific value?
5. Why are dermestid beetles of concern to insect collectors?

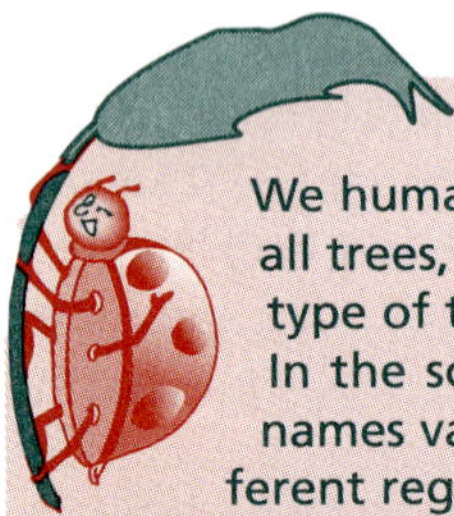

The Insect Naming Game

We humans like to name things. The use of language probably began with names: tree, dog, bug! But all trees, dogs and bugs aren't the same. So special names were necessary to distinguish between each type of tree, dog or bug.

In the scientific world, the names used in everyday language are called common names. Common names vary from place to place. So one animal might have several common names, each used in a different region.

Since most people don't use the Latin-based genus and species names, the common names for animals and plants still exist. The Entomological Society of America even publishes an "Approved Common Names List" for insects.

This list gives some insight into the basis for common names. For instance, color is used frequently in such names. There are 13 insects with green in their names, including a cloverworm, a scale, a budworm and a stink bug.

There are 19 insects with the color brown in their name; 22 if you count brown-headed, brown-banded and brown-legged. Black shows up in 39 insect names; 40 if you include the blackbellied clerid—a type of beetle. Yellow and white also are popular colors in insect common names.

Numbers are handy for common names, too. In insects, most numbers are used. There is the onespotted stinkbug. Two is a popular number as in twospotted, twobanded, twolined, and twostriped. There are threes and fours, but no fives, in insect names. There are sixes and even a sevenspotted lady beetle. There is an eightspotted forester, which just happens to be a moth. No nines show up in insect names. But there are two tens—a tenlined June beetle and a tenspotted lady beetle. There is also a thirteenspotted lady beetle.

Some common names reflect the plant the insect calls dinner. As you might guess, plants grown as crops occur more frequently in names than other plants. We have several insects with corn or bean in their names. There are 11 insects with spruce in their names, 10 with sunflower, eight with sweet potato and two with turnip. There is even a quar midge and a burdock borer.

Regions of the world are mentioned frequently in insect names. European, Asian and African each reflect the native area of the insect bearing the name. There is a Brazilian leafhopper and six insects with Australian in the name. Some U.S. states are used in insect common names. We have a Colorado potato beetle. And California has its name on nine insects.

Sometimes the insect name is directional as in western bean cutworm or southern corn rootworm. There are 44 names with western, 22 with southern, 15 with eastern and only eight with northern. Sometimes the name is combination directional, as with the southeastern blueberry bee.

Of course, the directional names don't mean much today. Take those pests of corn, the northern, western and southern corn rootworms. The range of the southern is further north than the northern, and the western is found further south than the southern! That hasn't always been the case, but common names aren't always up to date.

Classification, Nomenclature and Identification of Insects

Insects, like other animals, can be classified according to their phylogenetic (evolutionary history) relationships. This is the classification system used in biology today and it incorporates morphological (structure), physiological (processes of life) and ecological (relationship to environment) information about the organisms that are classified.

The classification system involves a hierarchy beginning with a broad category that includes one or more groups from the next lower level. This system is based on the system developed by Linnaeus, so it has been in use for more than 200 years. The hierarchy of categories used is as follows:

Kingdom
- Phylum
  - Class
    - Order
      - Family
        - Genus
          - Species

The use of this system to classify man and the housefly is illustrated in the table below.

The system of naming species (Nomenclature) also originated with Linnaeus. Each species is given a name consisting of two words (binomen)—the first the genus, the second the species.

So named, we humans become *Homo sapiens* and the house fly *Musca domestica*. The names are Latin or Latinized and are either *printed in italics* or underlined if handwritten. The first word in the binomial nomenclature (the genus) is always Capitalized, while the second (the species) is not.

Scientific Classification of the House Fly and the Human

Category	House Fly	Human
Kingdom	Animalia	Animalia
Phylum	Arthropoda	Chordata
Class	Insecta	Mammalia
Order	Diptera	Primates
Family	Muscidae	Hominidae
Genus	*Musca*	*Homo*
Species	*domestica*	*sapiens*

The same Latin scientific names are used throughout the world. In fact that is the only part of a Chinese scientific publication that I can understand! The other names that organisms acquire are called *common names*. Common names can vary and some organisms might have several within a country or even a state. For instance the word *gopher* is used in parts of Arkansas as a name for a kind of tortoise. In Minnesota what people call a gopher is a kind of ground squirrel. In many parts of the U.S. a gopher is a nearly blind, grey fuzzy creature that tunnels underground devouring insects and earthworms as it goes and all the while creating unsightly mounds of soil on lawns. If someone uses the name gopher it is not possible to visualize the creature being discussed without knowing where the person is from. However, all the animals called gophers have scientific names that (if you knew the scientific names) would leave no doubt in your mind what creature was mentioned. That is the purpose of scientific names—to increase ease of understanding and eliminate confusion in the world of science.

Insects can be identified through the use of a key. A commonly used type of key has bits of information arranged in pairs—called couplets. The user of such a key must decide which description in the couplet best fits the specimen to be identified. The key will direct the user to another couplet. If the proper decisions are made, the specimen will be identified by using such a procedure. Nowhere is the old saying "Practice makes perfect" more true than in using keys to identify insects.

Key to Insect Orders

A key to the adult stage of insects is provided on following pages. This key is designed to identify the insect as to order. Other keys are available that can be used to identify insects as to genus and species. In using this key the first decision to be made would be whether or not the specimen has wings. If so the user is directed to couplet 2, if not the direction is to couplet 28. The user of the key continues going through the couplets until an identification is made.

Positive insect identifications are almost always made by using a key. Identifications can be made without a key by using characteristics of the insect.

Key to Insect Orders

Key to the Adult Stage of Insect Orders

1. **(a)** Insect with wings 2
   (b) Insect without wings 28

2. **(a)** Front wings hardened, leathery or parchment-like, at least at the base 3
   (b) Front wings membranous 8

3. **(a)** Chewing mouthparts 4

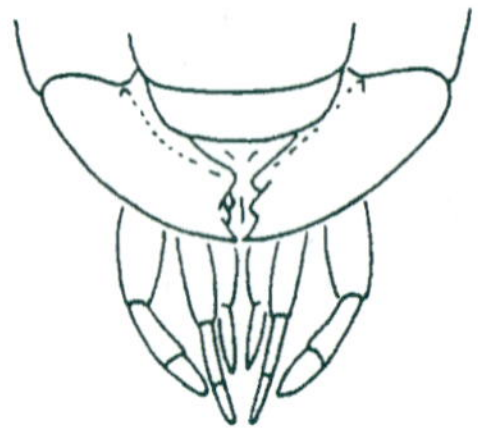

   (b) Sucking mouthparts 7

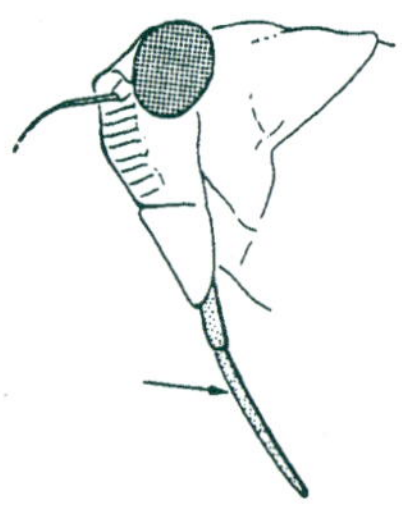

4. **(a)** Without pincer-like cerci 5
   (b) With pincer-like cerci (earwigs) .. DERMAPTERA

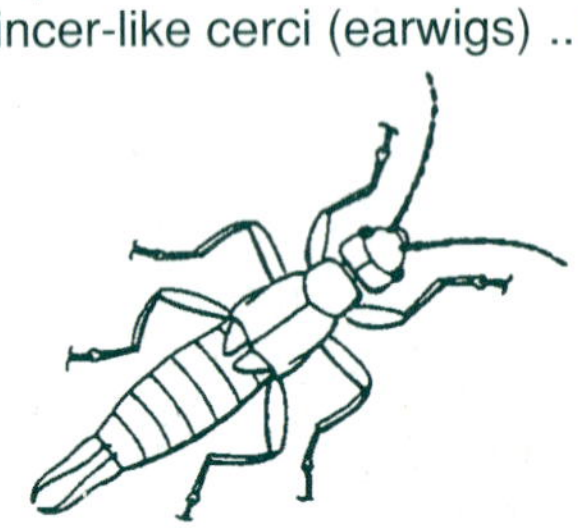

5. **(a)** Front wings with branched veins 6

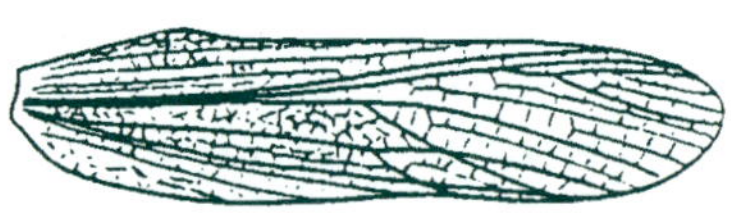

   (b) Front wings hard, without veins (beetles) .... COLEOPTERA

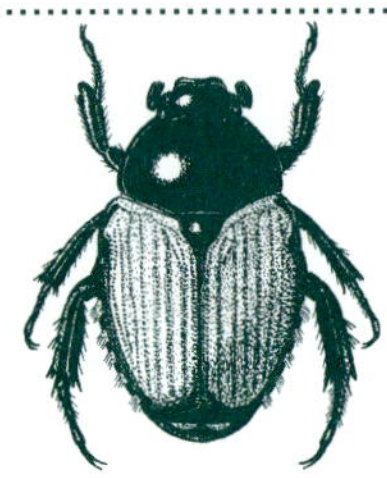

6. **(a)** Jumping insects; hind femur enlarged; tarsi with four or fewer segments (crickets, katydids, grasshoppers) .... ORTHOPTERA

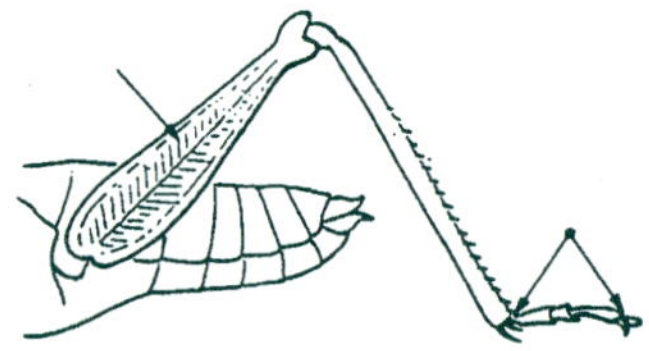

   (b) Walking insects; hind femur not enlarged; tarsi with five segments (roaches, mantids, walking sticks) .... DICTYOPTERA

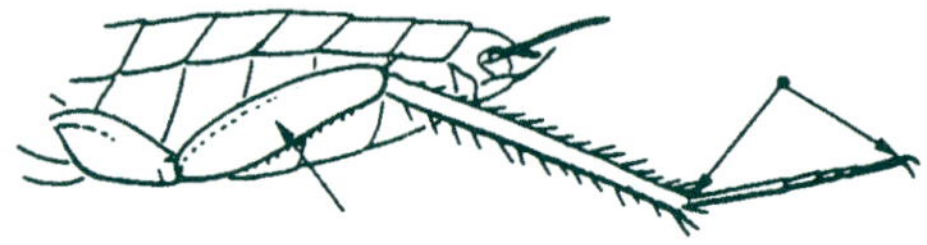

7. **(a)** Front wings leathery at base, membranous at end (true bugs) .... HEMIPTERA

   (b) Front wings of uniform texture (leafhoppers, treehoppers, spittlebugs, cicadas, aphids) .... HOMOPTERA

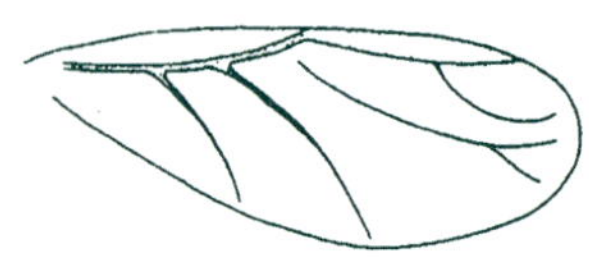

8. **(a)** Two wings 9
   (b) Four wings 13

9. **(a)** Pronotum extending over abdomen (pigmy grasshoppers) .... ORTHOPTERA

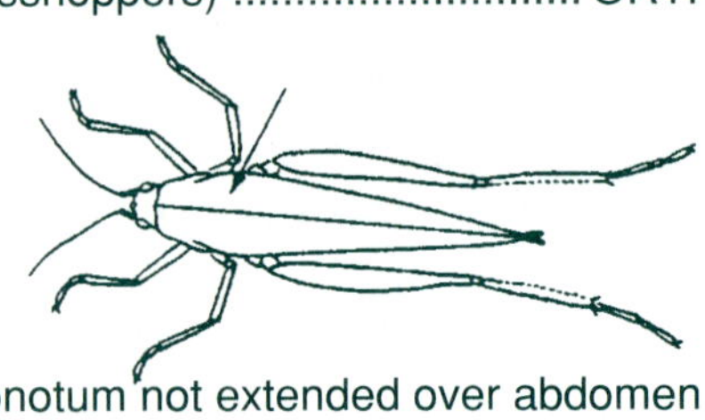

   (b) Pronotum not extended over abdomen 10

10. **(a)** End of abdomen without noticeable appendages 12
    (b) End of abdomen with style or thread-like tails . 11

→

11. (a) Style-like tail (male scales) .... HOMOPTERA

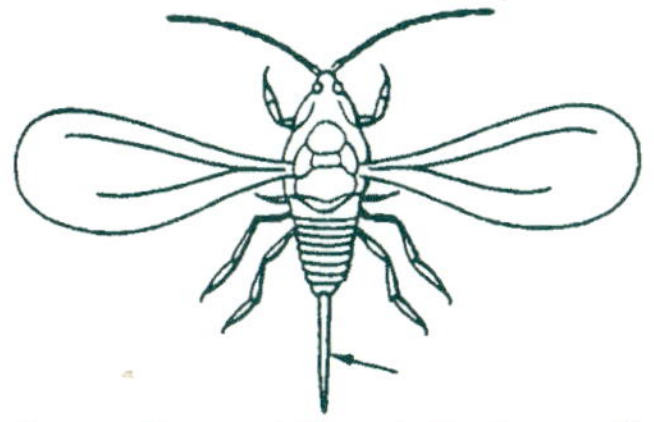

(b) Two or three thread-like tails (mayflies) .... EPHEMEROPTERA

12. (a) With haltere-like organs in front of wings (male stylopids) .... STREPSIPTERA

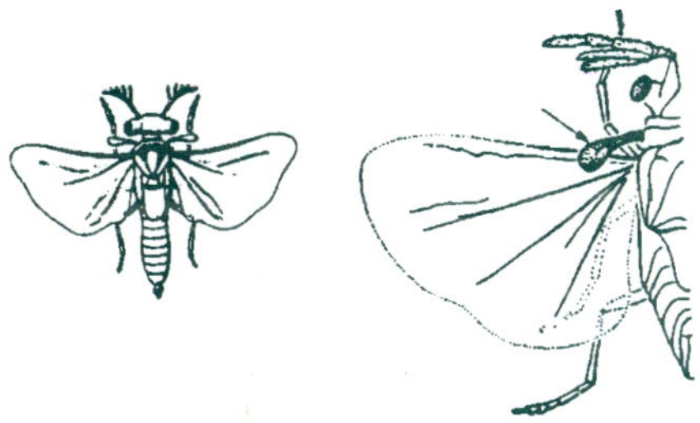

(b) With halteres behind wings (mosquitoes, flies, gnats, midges) .... DIPTERA

13. (a) Wings usually covered with scales; mouthparts consist of a coiled proboscis (butterflies, moths) .... LEPIDOPTERA

(b) Wings with few or no scales; without coiled proboscis 14

14. (a) Very slender wing with fringe of hairs as long as wing is wide (thrips) .... THYSANOPTERA

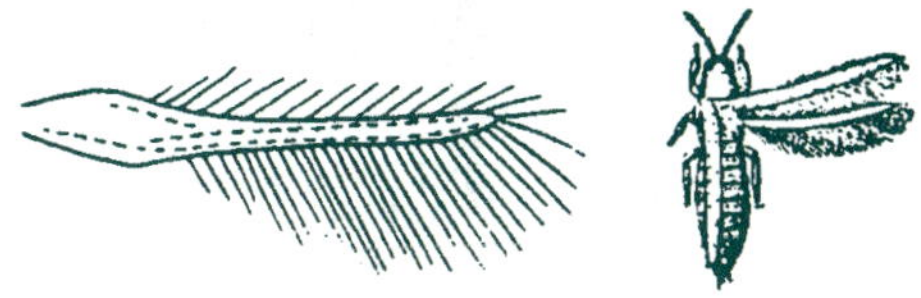

(b) No fringe of hairs, or if present, not as long as wing is wide 15

15. (a) Hind wings equal to, or larger than front wings 21

(b) Hind wings smaller than front wings 16

16. (a) No long abdominal tail-like appendages 17

(b) Abdomen with two or three thread-like tails (mayflies) .... EPHEMEROPTERA

17. (a) Tarsi two or three segmented 18

(b) Tarsi with more than three segments (usually 5) 20

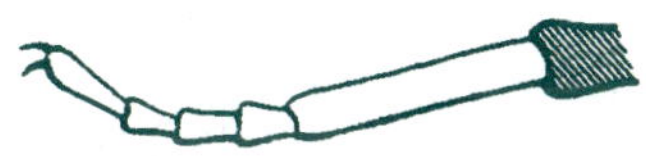

18. (a) Piercing-sucking mouthparts (leafhoppers, plant hoppers, spittlebugs, aphids, cicadas) .... HOMOPTERA

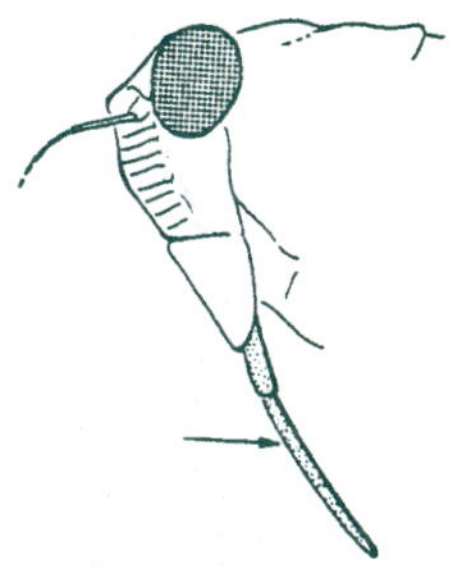

(b) Chewing mouthparts 19

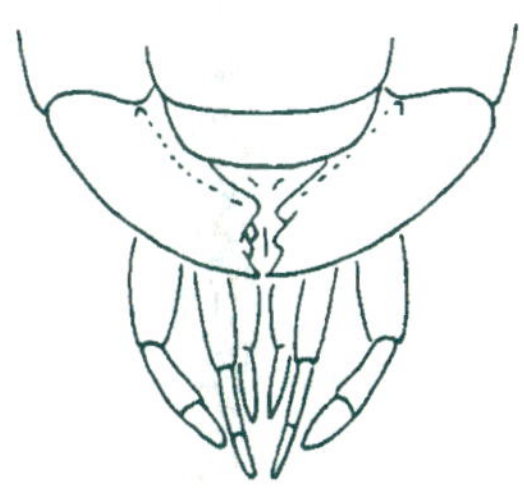

19. (a) Cerci present (zorapterans) .... ZORAPTERA

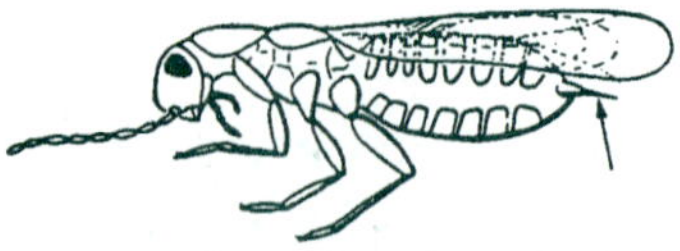

(b) Cerci absent (booklice, barklice) PSOCOPTERA

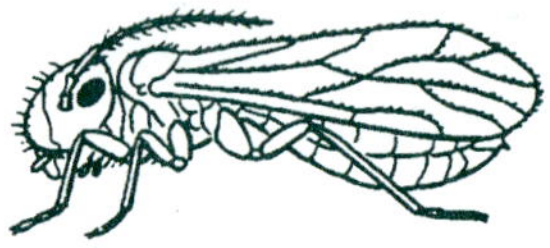

20. (a) Antennae shorter than body, not noticeable scales (bees, wasps, ichneumons) .... HYMENOPTERA

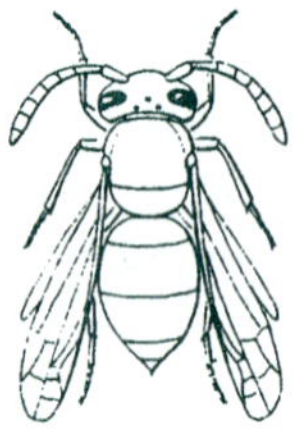

(b) Antennae as long as body; wings and body often with hairs (caddisflies) .... TRICHOPTERA

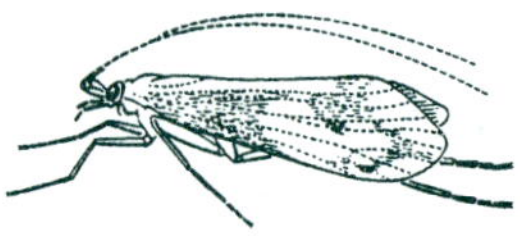

21. (a) Mouthparts close to eye 22

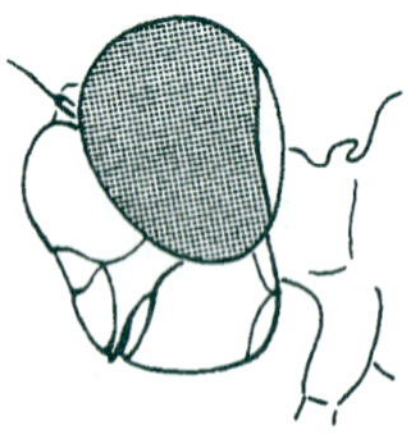

(b) Mouthparts at end of beak-like structure some distance from the eye (scorpionflies) .... MECOPTERA

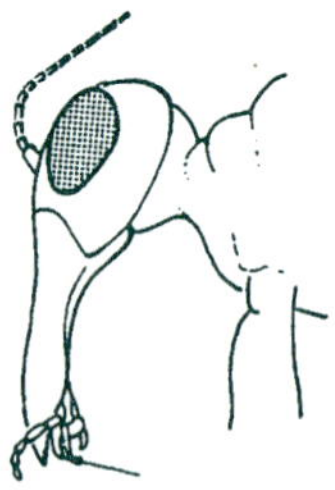

22. (a) Wings never held flat over abdomen 23

(b) Wings held flat over abdomen 25

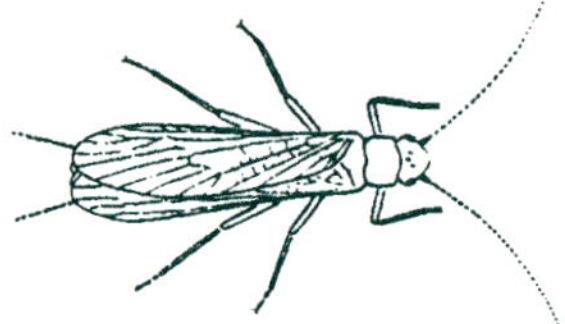

23. (a) Bristle-like inconspicuous antennae (dragonflies, damselflies) .... ODONATA

(b) Antennae apparently with several segments .. 24

24. (a) Hind wings with enlarged anal area folded fan-like; wings tend to curl around the body lengthwise (dobsonflies, fishflies, alderflies) .... MEGALOPTERA

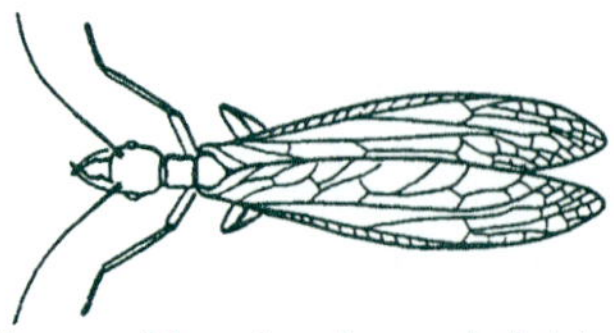

(b) Hind wings without enlarged, folded area; wings do not tend to curl around body lengthwise (lacewings, mantispids, ant lions, owl flies) .... NEUROPTERA

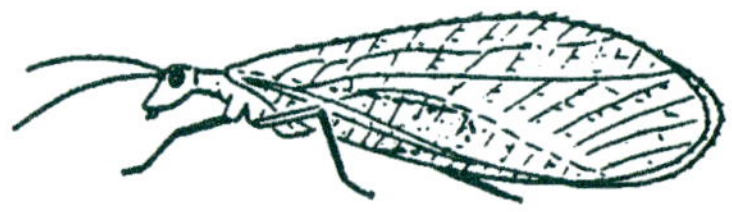

25. (a) All legs of walking type 26

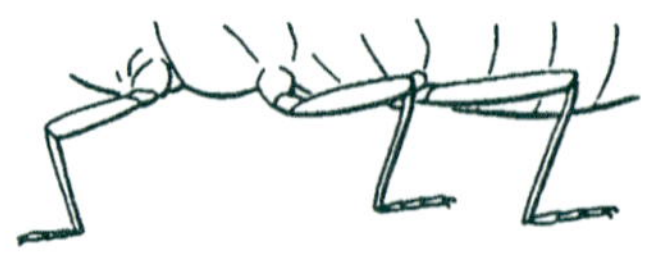

(b) Hind legs modified for jumping (tree crickets) .... ORTHOPTERA

26. (a) Basal segment of front tarsus has swollen appearance (webspinners) .... EMBIOPTERA

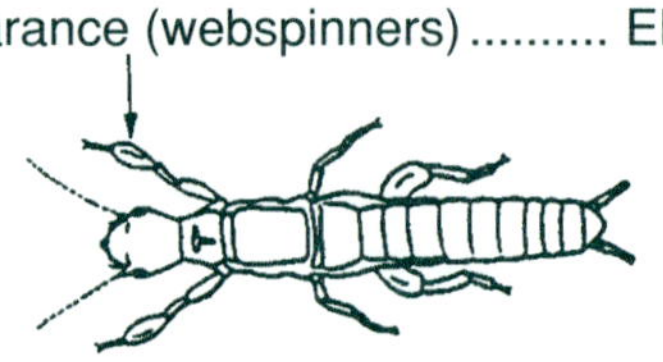

(b) All tarsal segments approximately equal in size 27

→

27. (a) Cerci usually long; more than 8 segments (stoneflies) .... PLECOPTERA

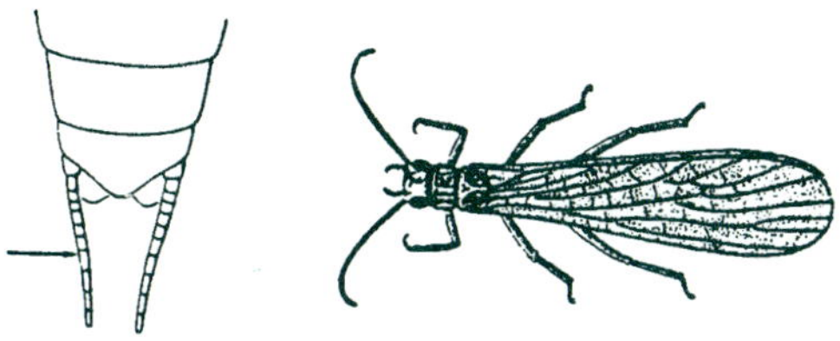

(b) Cerci short; with 2 to 8 segments or absent (termites) .... ISOPTERA

28. (a) Antennae present 29
(b) Antennae absent 40

29. (a) Mouthparts usually withdrawn or enclosed in the head and not apparent 30
(b) Mouthparts usually distinctly apparent 31

30. (a) Collophore present; spring-like organ usually present (springtails) .... COLLEMBOLA

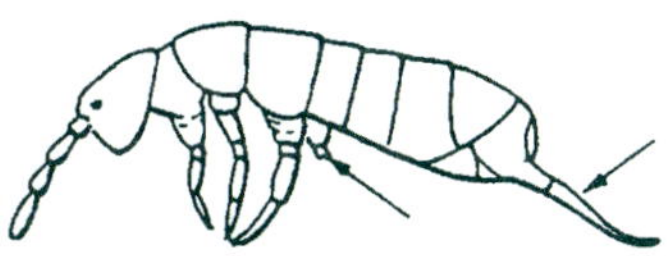

(b) Both collophore and spring-like organ absent; distinct cerci present .... DIPLURA

31. (a) Long tail-like appendages absent 32
(b) Three tail-like appendages present (silverfish, firebrats) .... THYSANURA

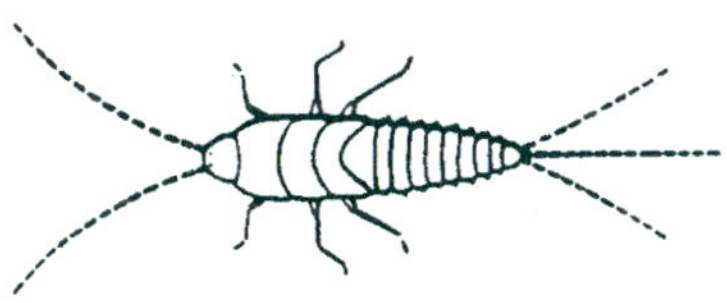

32. (a) Body flattened laterally or dorsoventrally 33

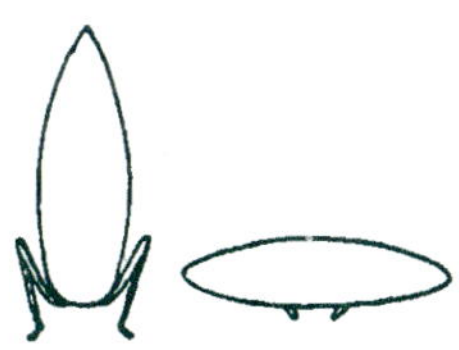

(b) Body not flattened 43

33. (a) Body flattened laterally (fleas) . SIPHONAPTERA

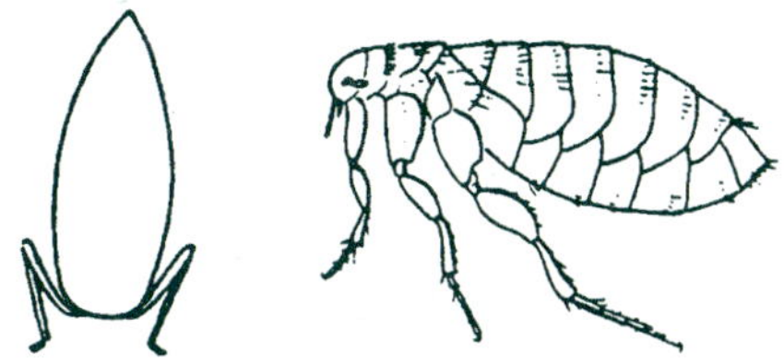

(b) Body flattened dorsoventrally 34

34. (a) Sucking mouthparts externally visible 35
(b) No sucking mouthparts externally visible 36

35. (a) Antennae longer than head (true bugs) .... HEMIPTERA

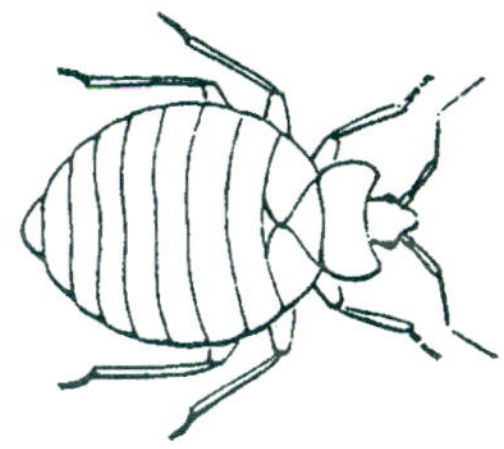

(b) Antennae shorter than head (louse flies, bat flies) .... DIPTERA

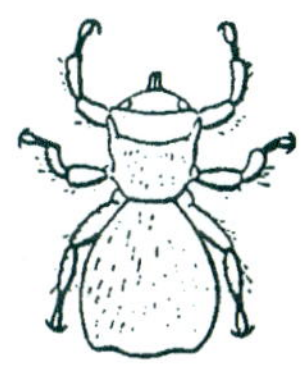

36. (a) Antennae longer than head 37
(b) Antennae shorter than head 39

37. (a) Basal segment of front tarsus swollen (web spinners) .... EMBIOPTERA

(b) Basal segment of front tarsus not swollen 38

→

38. (a) Tiny insects; tarsi 2 to 3 segmented (booklice, barklice) .... PSOCOPTERA

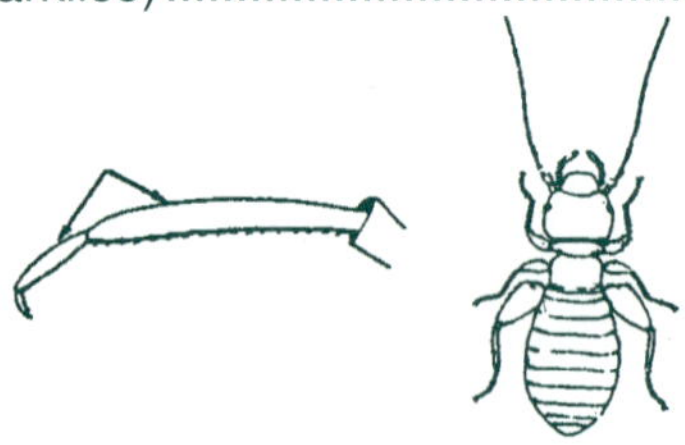

(b) Large insects; tarsi five segmented (cockroaches) .... DICTYOPTERA

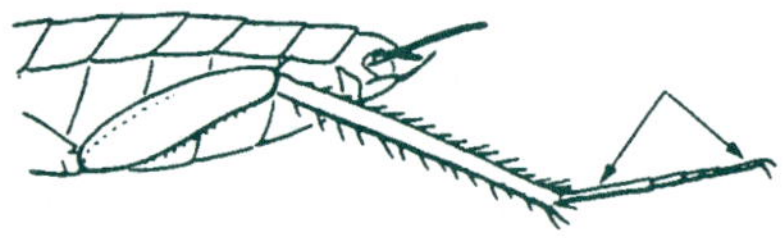

39. (a) Head wider than thorax at point of attachment to thorax (chewing lice) .... MALLOPHAGA

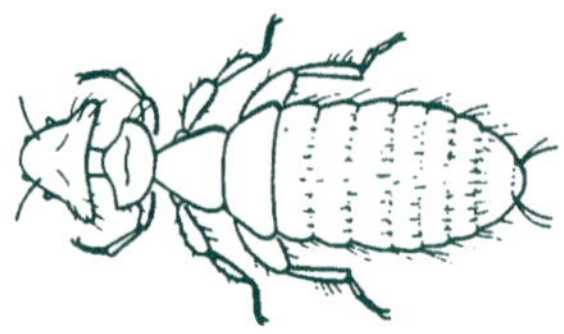

(b) Head narrower than thorax at point of attachment to thorax (sucking lice) .... ANOPLURA

40. (a) Legs present 41
(b) Legs absent.. 42

41. (a) Fewer than 12 abdominal segments (bat flies, louse flies) .... DIPTERA

(b) Twelve abdominal segments (proturans) .... PROTURA

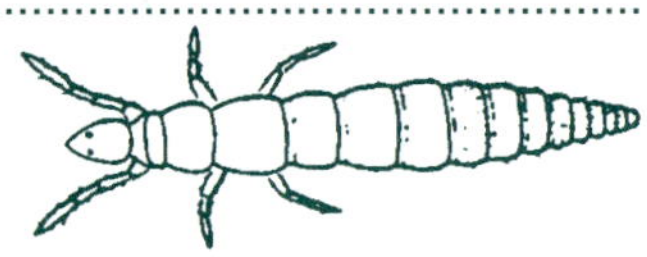

42. (a) Head and thorax separate (scale insects).......... ....HOMOPTERA

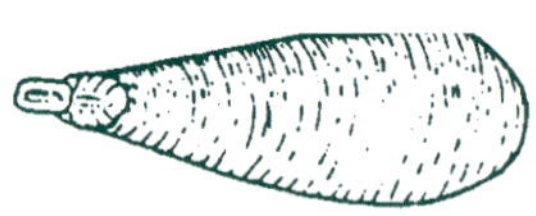

(b) Head and thorax fused (female stylopids).......... .... STREPISIPTERA

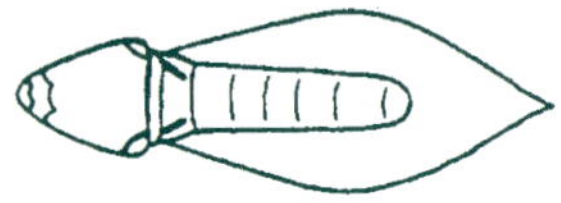

43. (a) Abdomen and thorax narrowly joined together (ants) .... HYMENOPTERA

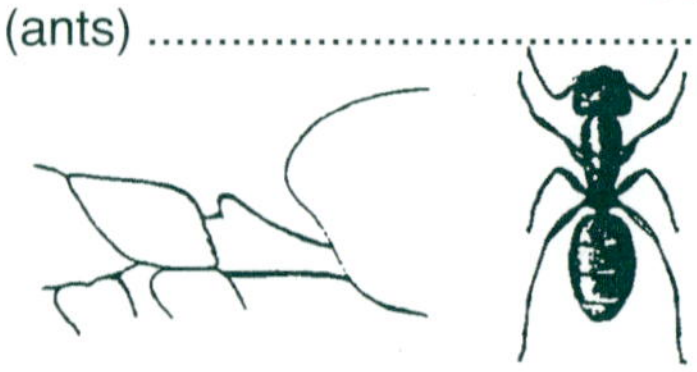

(b) Abdomen and thorax broadly joined together 44

44. (a) Body covered with scales (female cankerworms) ....LEPIDOPTERA

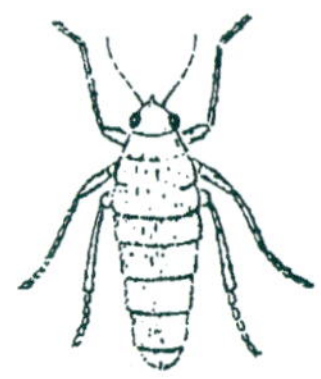

(b) Body not covered with scales.......................... 45

45. (a) Tarsal claws absent (thrips) .. THYSANOPTERA

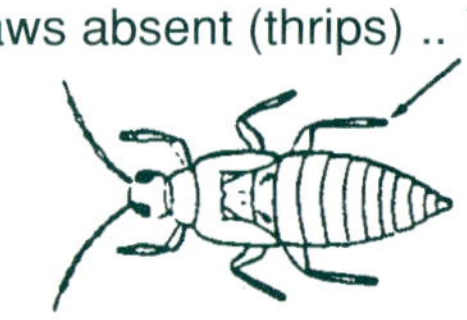

(b) Tarsal claws present 46

46. (a) Piercing-sucking mouthparts 47
(b) Chewing mouthparts 49

47. (a) Cornicles usually present (aphids or plant lice)... ....HOMOPTERA

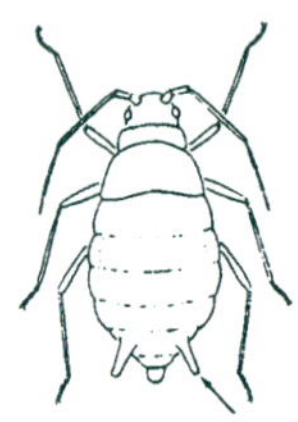

(b) Cornicles absent 48

→

48. (a) With distinct head and eyes (bed bugs) .... HEMIPTERA

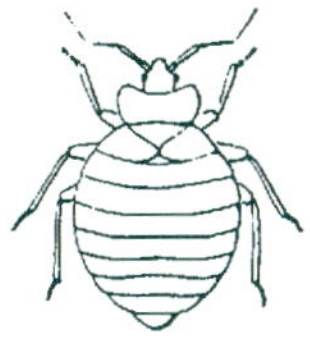

(b) Without distinct head and eyes (female scales) .. .. HOMOPTERA

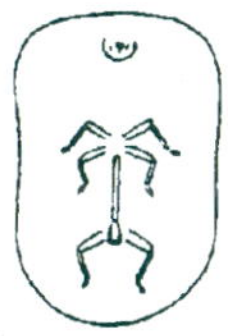

49. (a) Abdominal forceps present (earwigs) .... DERMAPTERA

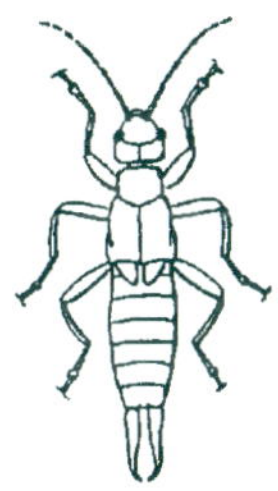

(b) Abdominal forceps absent 50

50. (a) Mouthparts at end of beak like structure some distance from eye (scorpionflies).. MECOPTERA

(b) Mouthparts not elongated, close to eyes......... 51

51. (a) Cerci present 53

(b) Cerci absent 52

52. (a) Antennae longer than one-third of body length (booklice, barklice) .... PSOCOPTERA

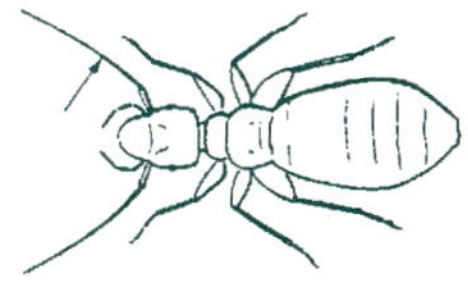

(b) Antennae shorter than one-fourth of body length (female stylopids) .... STREPSIPTERA

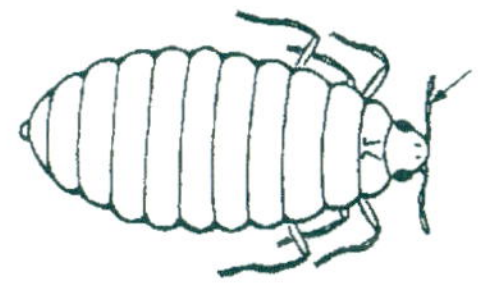

53. (a) Body leathery and usually grey or dark colored 54

(b) Body soft and pale colored 57

54. (a) Hind legs adapted for jumping .... ORTHOPTERA

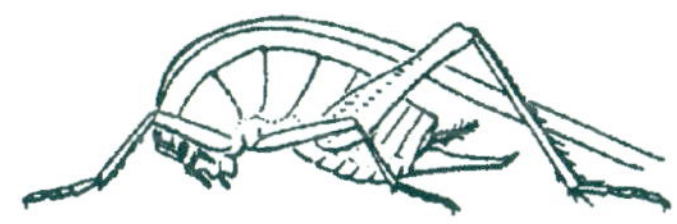

(b) Hind legs not adapted for jumping 55

55. (a) Pronotum narrow, body pencil-like or stick-like ... . PHASMIDA

(b) Body not elongated and stick-like 56

56. (a) Front legs enlarged grasping type (mantid) .... DICTYOPTERA

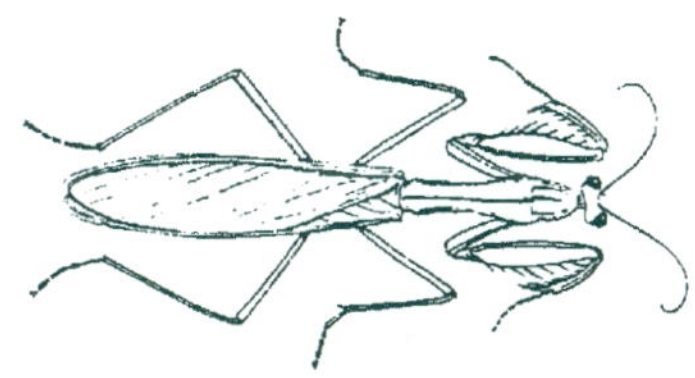

(b) Front legs not enlarged, pronotum broad and flattened, frequently pronotum covers head (cockroach) .... DICTYOPTERA

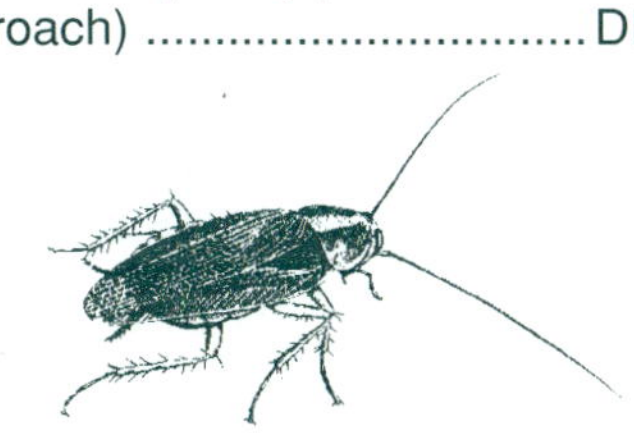

→

57. (a) Two tarsal segments (zorapterans) .... ZORAPTERA

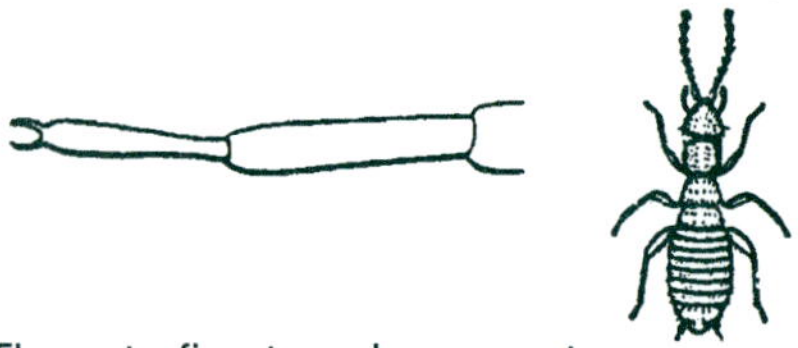

(b) Three to five tarsal segments 58

58. (a) Basal segment of front tarsai with swollen appearance (webspinners) .... EMBIOPTERA

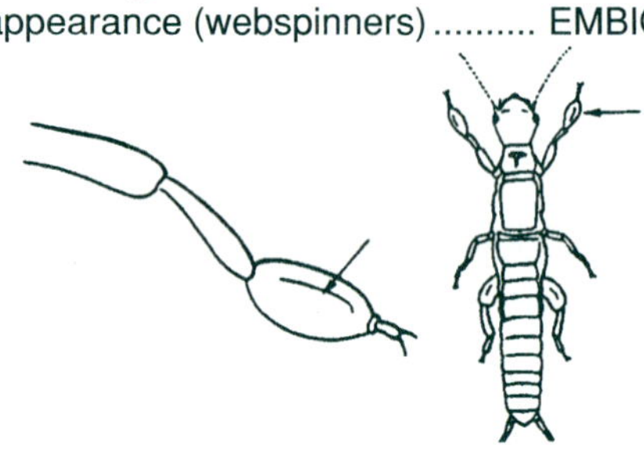

(b) Basal segment of front tarsi about same size as ones immediately following (termites) .... ISOPTERA

Such sight identification can be used only by learning what the insect looks like. However, many insects look alike and inexperienced (and sometimes experienced) insect identifiers can make mistakes by using sight identification. This is especially true if the identification needs to be at genus and species levels.

References

Blackwelder, R.E. 1967. Taxonomy; a text and reference book. J. Wiley and Sons, New York. 698 pp.

Borror, D.J., D.M. DeLong, and C.A. Triplehorn. 1981. An introduction to the study of insects. Saunders College Publishing, Chicago. 827 pp.

Brues, C.T., A.L. Melander, and F.M. Carpenter. 1954. Classification of insects. Harvard Univ. Mus. Compar. Zool. Bull. 73. 917 pp.

Comstock, J.H. 1964. An introduction to entomology. Comstock Publ. Assoc., Ithaca. 1064 pp.

Imms, A.D. 1957. A general textbook of entomology. Methuen and Co., London. 886 pp.

McCafferty, W.P. 1981. Aquatic entomology. Sci. Books Internat., Inc., Boston. 448 pp.

Usinger, R.L., Ed. 1956. Aquatic insects of California with keys to North American genera and California species. Univ. of California Press, Berkeley. 508 pp.

Joint-Footed Creatures

Biologists classify all living things into groups. The broadest classification is either plant or animal. These groups are called kingdoms and each kingdom is divided into subgroups called phyla.

One such phylum of the animal kingdom is Arthropoda. The word arthropod is based upon the Greek term "arthr," which means joint and "pod," which means a foot. We find the same combining of terms in arthritis, an inflammation of a joint, and podiatrist, a foot doctor.

Arthropods are nature's joint-footed animals that possess a hard outer skeleton. The most prominent of the arthropods are the insects. Insects have three body regions: the head, thorax, and abdomen. They also have six legs, two antennae and most have wings.

Crustaceans are also arthropods and include the water dwelling crabs, lobsters, crayfish and shrimp. Most have four antennae and appendages on segments of the abdomen.

Some arthropods are identified because they have many legs. The slow-moving millipedes are worm-like creatures that have four legs on each body segment. They don't have to be fast afoot because they feed on dead organic matter. Pillbugs, sometimes called sowbugs or roly-polies, are a type of millipede. Centipedes, on the other hand, move very fast. They are predators and must catch up to the other arthropods on which they feed. Centipedes have only two legs on each body segment.

The last major group of the arthropods is Arachnida. Arachnids include the spiders, granddaddy longlegs, ticks and mites. All have eight legs, no antennae or wings, and only two body regions. Because of this, some folks say that spiders appear to have lost their heads. Spiders and granddaddy longlegs, also called harvestmen, are predators of other arthropods.

So there you have it, the family of insects. You might say that mites, ticks, crayfish, millipedes, centipedes and insects are cousins. Some people don't worry about the relationship and lump all of the arthropods together under one heading. They call all of nature's joint-footed creatures "creepy crawlies!" And I don't think it could be considered a term of endearment.

Insect Wings Useful to Insects and Entomologists

One of the outstanding attributes of adult insects is that they have wings. Wings allow insects to travel to new places in search of food or mates. The ability to fly also provides a mechanism to escape enemies and to find a place to spend the winter. Wings are one of the keys to success for this group of animals.

But not all insects have wings. There are some ancient groups of insects that are wingless. For instance, those insects called firebrats and silverfish lack wings. The same is true of collembola, tiny insects that feed on organic matter in the soil.

But most insect species have wings in the adult stage. On the other hand, immature insects do not have wings. But in that stage, the lack of wings is a benefit since many immature insects feed in confined spaces, and wings would just get in the way. Such is the case with fly maggots, which feed in dead animals or garbage heaps, the grubs that crawl around in the soil looking for roots, and some caterpillars that literally bore through the plant tissue on which they feed.

In some insects, such as termites and ants, the wings are present in reproductive adults but not in other forms. Like many immature insects, the workers of these social insects crawl around in the tunnels of the nest where wings would just get in the way. Queens of ants and termites actually chew off their wings, following mating flights.

Wings are very handy to insects as tools for locomotion. Insect wings also are very handy to entomologists who classify insects. Entomologists frequently use the variation of wing structure of insects to write descriptions of species.

Most of the order names of insects are based on a Latin root for the type of wing possessed by that order. Many of the insect order names include "pteron," Latin for fur, wing or feather. In the case of insects, it means wing.

Flies are classified in the order Diptera. "Ptera" for wing and "di" for two, in reference to the two wings possessed by flies. Two wings are unusual for insects, since other winged forms have four wings.

Beetles have a hard pair of outer wings called elytra. Coleos, the Latin term for sheath, is the basis for Coleoptera, the order name for the beetles. Hemi means half; consequently, the insects in the order Hemiptera are half-winged. These insects have a leathery front half of the wing while the back portion is membranous.

Butterflies and moths have scales on their wings. So these insects are classified in the order Lepidoptera, which interpreted literally means "scale wing." We find the same Latin root in the human disease leprosy where the skin tends to flake or scale off. Even the poet Robert Frost noted the colored scales on the wings of butterflies and used the term "dye-dusty wing" to describe the wings of monarch butterflies in his poem, "Pod of the Milkweed."

Lacewings fall in the order called Neuroptera because their wings appear to be traversed by nerves. Grasshoppers and crickets are Orthoptera, due to the presence of straight outer wings. The same Latin root is found in the profession orthodontist, the dental specialist who straightens our teeth.

It is obvious that wings are important to insects as a method of locomotion. It's also true that these wings provide obvious characteristics to help scientists separate insects into groups called orders. For both insects and entomologists, the wings have it!

What's In a Name?

Entomologists, like other scientists, name living things according to a two-name system called binomial nomenclature. For instance, in this scientific lingo humans are called *Homo sapiens,* our official genus and species. Scientific names are Greek to some people. And for good reason. Most scientific names are based on Greek and Latin roots.

Scientific names serve a very useful purpose. They are universally recognized across all languages. However, most people do not use scientific nomenclature in their daily language. As a result, most living things are known by common names. So it is with the insects.

What's in a name? Some say quite a bit about their owners. Insect common names sometimes reflect a geographical location. We have Chinese mantids, Carolina grasshoppers, California harvester ants, European hornets, German cockroaches, Hawaiian carpenter ants, and Tahitian coconut weevils.

Some names are such that you might want to use them to describe your worst enemy. For instance, tropical rat louse, western treehole mosquito, twig pruner, vagabond crambus, uglynest caterpillar, goat sucking louse or even corn sap beetle.

Names sometimes reflect the food that an insect uses. We have coconut scales, corn leaf aphids, cranberry weevils, onion maggots, yucca moths, raspberry sawflies, and cotton borers.

A few insects lay claim to noble birth. We have monarch butterflies and regal moths. Other insects are of more humble ancestry. Those insects include the common cattle grub, the common green lacewing and the common malaria mosquito.

Some insects apparently behave in ways that entomologists thought unusual. The crazy ant, the confused flour beetle, the gloomy and greedy scales, and the thief ant all fall into this category.

Other insects, it appears, could use some grooming tips. The crinkled flannel moth might need an iron; a shower could benefit the rough stink bug; and perhaps the roughskinned cutworm just needs a little lotion. The resplendent shield bearer, however, is fine as it is.

If imitation is the sincerest form of flattery, then toad bugs, rhinoceros beetles and crab lice have paid their compliments. In addition to borrowing their names, they also bear some resemblance to their namesakes.

Some insects are small, like the small milkweed bug. Others are smaller, such as the smaller yellow ant. As might be suspected, we also have a large milkweed bug and a larger yellow ant.

If all of this seems a bit confusing, well it is. You see, that is the purpose for those scientific names. For instance, take that famous insect called the small pigeon louse. It's not really clear if it is a small louse that attacks pigeons or if it is a louse that attacks small pigeons.

No matter; to clear up the confusion, it has a scientific name: *Campanulotes bidentatus compar.* Isn't science wonderful?

A Bee or Not a Bee?

Millions of insects inhabit planet Earth. To keep track of these six-legged creatures, scientists divide insects into groupings called orders. Beetles belong to the order Coleoptera. The order Orthoptera includes grasshoppers and crickets, while bees, ants and wasps are Hymenoptera.

Nearly 900,000 insects have been given a scientific name. Each scientific name includes at least two parts: a genus and a species name. The genus corresponds to a human family name, and the species is equivalent to a human first or given name. This method of naming organisms is known as bionomial nomenclature—a two-name system.

Scientific names are normally Latinized words, which makes them somewhat cumbersome for use by the general public. For instance, everyone would recognize the *Musca domestica*, but almost everyone calls that insect by its common name: house fly.

But even insect common names can be a bit confusing. For example, many insects fly, but all insects that fly are not flies—insects technically of the order Diptera. Thus, the common dragonfly is not a fly at all, even though it is an accomplished flyer.

To avoid this confusion when using common names, most entomologists use the following rule: If an insect is not technically a member of the order for which it is named, then its common name is written as one word.

Take, for instance, our friend, the dragonfly. Also the dobsonfly, damselfly and lanternfly. True flies, such as the house fly, the horse fly, the deer fly and the bot fly, have their common names written—or, rather, written properly—as two words. (Some dictionaries disagree, leaving them etymologically correct, perhaps, but not entomologically correct!)

The same is true for insects called bugs. All bugs—members of the insect order Hemiptera—are insects, but all insects are not bugs. For example, the ladybug is not a bug at all but a beetle. So ladybug should be written as one word. If, however, this insect is called a lady beetle, two words are correct. The same is true of the insect called a Junebug; it's really a beetle and is sometimes more correctly called a May beetle. The lightningbug is frequently called a firefly, but it is neither a bug nor a fly, it is a beetle.

Some insects look like insects of another order, but their common names—when written in proper entomological fashion—can betray their charade. To wit, the bee fly and the bee moth. These imposters should not be confused with the real McCoy, such as the honey bee and the bumble bee.

So, if the question is, "A bee or not a bee?" the answer would be found in the way it's written.

Classification, Nomenclature and Identification of Insects

1. What is meant by classification according to phylogenetic relationships?
2. What classification category do humans and insects share?
3. What is binomial nomenclature?
4. What are common names of animals?
5. What is the correct method of writing genus and species names?
6. How can you tell if an insect with the name fly in its name is a true fly (a member of the insect order Diptera)?

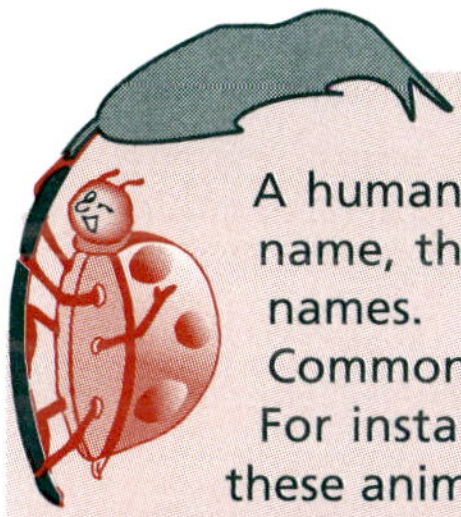

Insects Named for People?

A human endeavor of long standing is to name things. In biology a living organism gets a scientific name, the genus and species. But plants and animals, including insects, also are known by common names.

Common names can vary from locality to locality and might reflect characteristics of the organism. For instance, the whitetail deer, cottontail rabbit and bald-faced hornet have names descriptive of these animals.

One approach to creating scientific names for animals is to name the creature for an individual. For instance, one of a group of birds known as thrashers is *Toxostoma lecontei.* Named for J.L. LeConte, the man who discovered the bird, it is sometimes called the desert thrasher or LeConte's thrasher.

In his book *Pioneer Naturalists*, Howard Ensign Evans points out that entomologists are especially fond of naming insects for colleagues or people they admire. Most insects that have common names are pest insects since these are the insects humans most often encounter. It would seem that having an insect pest named after you would be a dubious honor. However, entomologists seem not to mind.

Pseudococcus comstocki is a notorious pest insect. It is a mealybug, a small wax-covered insect that sucks plant juices. John Henry Comstock was a distinguished entomologist well known for his work on mealybugs, so he was no doubt pleased to have an insect known as Comstock's Mealybug.

Several insects are named after people. Uhler's Assassin Bug, a small insect that feeds on other arthropods, is named after Philip Reese Uhler, who in 1864 was the first person to collect insect specimens and place them in a museum.

A beautiful sphinx moth is called Abbot's Sphinx. It's named after John Abbot, a watercolor painter who collected and sold insects. He also has a less attractive insect, a bagworm, named after him.

People other than entomologists also have insects named after them. For instance, Ibsen's wasp is a very small insect, less than 2 millimeters long, from the group of small wasps known as chalcids. It was named, for unexplained reasons, after the Norwegian playwright Henrik Ibsen by a chalcid specialist, Girault. We must assume that Girault was a fan of Ibsen.

Members of royal families sometimes are honored by having living things named after them. A common wild carrot is known as Queen Anne's Lace, supposedly because Queen Anne of England challenged her ladies-in-waiting to model their embroidery after the flower of this roadside weed.

Queen Alexandra, wife of King Edward VII of England, has two butterflies named after her. The first is Queen Alexandra's Sulfur Butterfly, which is a close relative of the pests known as cabbage and alfalfa butterflies. The butterfly just happened to be described in the year that Princess Alexandra married Albert Edward, son of Queen Victoria. Alexandra also has the largest butterfly in the world named after her—Queen Alexandra's birdwing. It just seems the right thing to do to name the largest butterfly in the world after the queen, especially if she also has a pest butterfly named after her.

UNIT 2

The Insect as a Life Form

CHAPTER 4

Insect Structure and Function

Adult insects typically have three distinct parts to the body. The parts are the head, thorax and abdomen. The insect body is basically composed of 20 segments although some of the segments are fused and it is not possible to count them. The head is composed of six segments. There are three segments in the thorax and 11 in the abdomen. Each of the parts of the insect body is adorned with protrusions—wings, legs and other things.

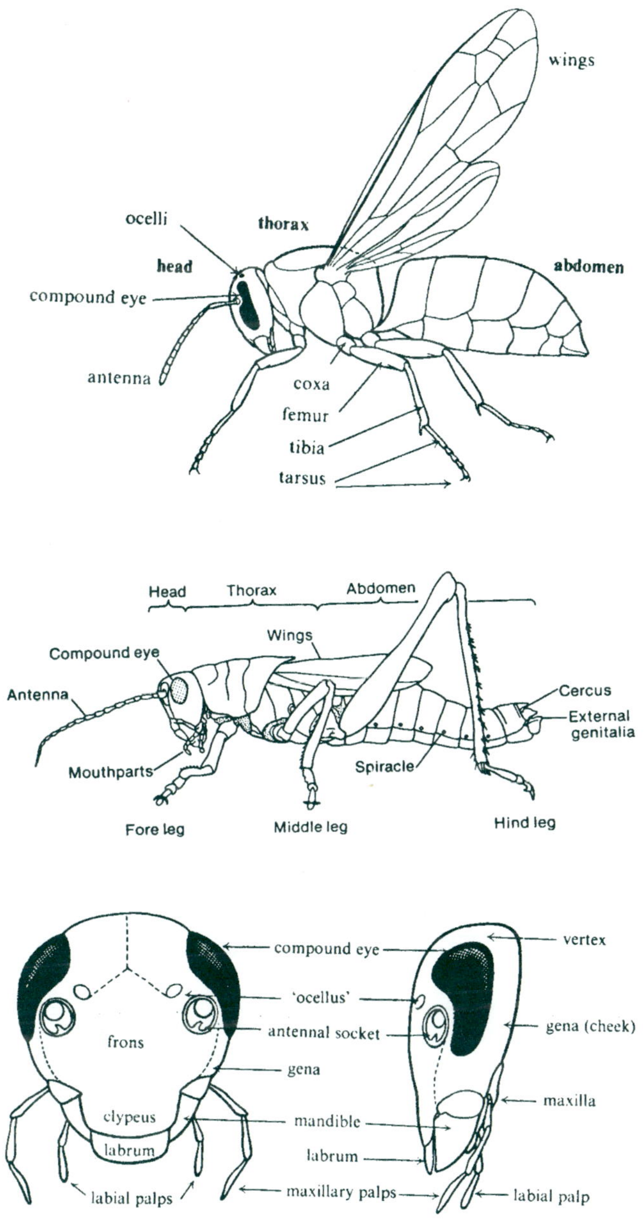

FIGURE 4.1. *Insect Structure.*

Head

The head of the insect functions primarily in a **food acquisition** and **sensory** role. The insect head is adorned with a pair of **compound eyes** (compound because they have more than one lens—the better to see you move) and a pair of **ocelli** (eye spots); two **antennae**—to feel and smell; and a mouth. The mouth, in a typical chewing insect (cockroaches, crickets and grasshoppers), is a box-like structure with a lid called a labrum. On each side of the mouth are two chewing devices: the mandible and the maxilla. The maxilla typically has an antenna-like structure called the palp. The floor of the mouth is the labium which may also have a pair of palps. The structures forming insect mouths are believed to have developed from the legs on the segments of some insect ancestor that looked a bit like a centipede with a pair of legs on each segment.

The mouth structure varies widely among insects. In houseflies the mouth is a sponging-sucking apparatus. Mosquitoes and bugs are equipped with piercing-sucking mouthparts. Butterflies and moths have a coiled tube for sipping nectar. Honey bees have chewing-lapping mouths.

Thorax

The middle part of the insect body is specialized for **locomotion**. On the thorax we find the wings and legs. There are three pairs of legs, two on each segment of the thorax. Like the mouthparts the legs differ among insects. They can function for running

In Insects, the Eyes Have It!

Nothing is more important to adult insects than their eyes. Looking casually at insects, most humans are more likely to notice other appendages first—legs, wings and antennae, for instance. But to the insect, the eyes are just as important as these more obvious protrusions.

To be sure, legs allow insects to run, jump and cling. Their wings provide the power of flight. Antennae are wonderful devices for smelling. But the eyes play very important roles in many aspects of insect lives.

Some insects, those like lice and fleas that are parasitic, don't have eyes at all. A few insects have single-lens eyes, but most adult insects have compound eyes. In some insects, the number of lenses is in the thousands. In dragonflies, the number of lenses is more than 10,000.

Because of compound eyes, entomologists assume that insects don't "see" in the same way that humans do. A single lens in an eye allows a sharply defined image. The multiple lenses in compound eyes probably, at best, produce a blurred image.

But then, the goal of the insect eye is not to provide the clearest image but to keep in visual contact with a large area. That is the reason that eyes sometimes occupy a sizable portion of the insect head. The eyes also frequently bulge out to develop a wide field of vision.

In some insects, such as dragonflies and some flies, the eyes actually meet in the midline of the head. Through multiple lenses, these insect eyes function much as a radar system that continually rotates, scanning the horizon. And the purpose is much the same: to pick up anything within sight.

While insect eyes might not provide the clearest image, they do provide an excellent motion detector. That is the reason it is difficult to sneak up on an insect. Insects need such protection, since it seems the world is out to get them—primarily to make a meal out of them. This is the case in insect eaters, such as birds and frogs.

Humans sometimes just want to remove the insect from the face of the earth. We developed such things as fly swatters to aid us in the task. But you still have to be quick, since it is a sure bet that the fly sees the swatter coming and many times manages to escape.

Insects also use their eyes to judge distance. This is important to predaceous insects that need to catch other insects for their food. The praying mantid uses it eyes for that purpose. By looking at a potential insect meal, it is able to accurately assess the distance to the prey, based on the view from several different lenses. It, therefore, waits until the prey is within range before grabbing for it.

Insect eyes are also sensitive to UV light waves. Such sensitivity allows the insect to see things that we humans cannot. Honey bees detect the UV patterns on flowers. Night-flying insects navigate using UV light. That is the reason that UV traps are so effective in luring insects to their deaths in an electric grid.

Polarized light is also detected by insects. Honey bees navigate based on the polarized light from the sky. This is known as the light-compass reaction where the insect can detect the pattern of polarized light from a patch of blue sky.

So to an insect, "I can see clearly now" doesn't have much meaning. Insect eyes are more UV light and movement sensors. But the system seems to work, since insects don't have trouble finding their way or avoiding predators.

Insect Mouths Have a Story to Tell

An old adage holds that we are what we eat. And as my mother often reminded me in my younger days, how we eat also speaks volumes.

Insects don't have mothers who are obligated to suggest the importance of table manners when they eat. But how insects eat does tell a story. The structure of insect mouths is related to their food and how they consume it.

An insect's mouth varies in structure from species to species and is one of the characteristics that scientists use to classify insects into groups. Most insect mouths can be categorized as one of six major types.

First, there is the chewing type. This is the type of mouth that scientists believe was the most primitive of insect mouths. Chewing insects have mandibles that cut off and grind the food—just like humans have mandibles for the same purpose. Many common insects possess chewing type mouths, including grasshoppers and crickets, cockroaches, praying mantids, beetles, dragonflies and most wasps. Caterpillars, the immature forms of butterflies and moths, also have chewing mouth parts.

Many of the insects with chewing mouths, such as grasshoppers, beetles and caterpillars, feed on living plants. Crickets, cockroaches and some beetles have to chew up dead stuff for food. Praying mantids, dragonflies, many wasps and some beetles are predators and must chew up their insect prey.

Other insects, like the non-biting flies, have a sponging type mouth. Like the house fly, these insects can eat only liquid food. Sometimes, if the item is a solid, the insect will vomit saliva on the potential food to turn it into a liquid. The liquid can then be sponged up.

Some biting flies, like horse and deer flies, have a cutting, sponging type of mouth. These insects cut the host's skin to cause blood to flow. The blood is then sponged up with the mouth parts, much the same way a surgical team would sponge up blood during an operation.

Many insects consume juices as food. These insects frequently have what are called piercing-sucking mouths. They have a hollow needle that is used to pierce plants in order to drink sap. Aphids, leafhoppers and cicadas feed in this way. Some insects use the same technique to drink blood from their prey. Mosquitoes, lice, fleas and bedbugs sup in this way.

Other insects also feed on liquids, but only if the liquids are free for the taking. Butterflies and moths have a siphoning tube for this function. The tube is called a proboscis and is generally curled up under the head of the insect when not in use. When the butterfly wants a sip of nectar, it uncoils the tube to produce a straw for sipping nectar from the flower. They also drink water in the same way.

Honey bees and bumble bees have a combination of chewing and drinking mouths. Called chewing-lapping mouths, they can drink nectar, as well as chew holes in most materials. That is how they sip nectar from a flower and chew up the wax to form the comb in the hive.

So like humans, how an insect eats sends a message. The chewers, like the caterpillars, chomp and rip their food with little thought to manners. On the other hand, butterflies sip their food in such a prim and proper way that even "Miss Manners" would be proud. Who would know that based on their eating habits the caterpillar could ever turn into a refined butterfly? Human moms take heart! There is hope for that chomping and chewing teenage son in the house.

(cockroaches), jumping (crickets), holding onto hairs (lice), swimming (water bugs), grasping (praying mantids) or digging (mole crickets). Some insect legs may be modified to provide a device for noise-making as is the case in some grasshoppers.

Wings are very important to insects as a method of locomotion. Wings are also widely used by entomologists as a method of classifying insects. Many insect orders can be based on wing characteristics. The Greek word "**pterum**" means a wing, feather or fin. That Greek word has become the root of many insect orders. For example Hemi**ptera**, the true bugs, which literally means half-wing, a reference to the fact that the wing is half leathery and half membranous. Not all adult insects are winged. For example, lice, fleas, and some ants are wingless. Wingless insects are called apterous—a (without) and ptera (wings). Most winged insects have four wings—two on each of the second and third segments of the thorax. However, the flies (Di**ptera**) have only two (*di* = two) wings—the back pair having been modified into balancing organs called halteres.

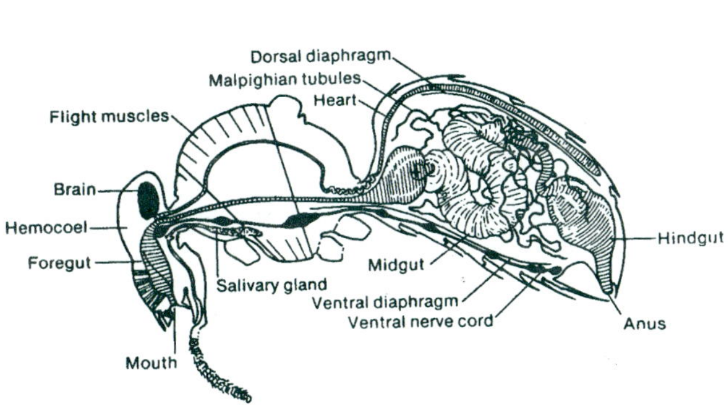

FIGURE 4.2. *Anatomy of an Insect.*

Insect wings are not thought to be modified legs because each segment of the thorax already has a pair of legs. Wings seem to have developed separately as extensions of the cuticle. Wings, like legs, are greatly diversified among the insects. Some wings are shell-like (beetles), others are covered with scales (butterflies and moths), or leathery (grasshoppers and cockroaches) or clear (cicadas).

Abdomen

The function of the abdomen is **breathing, excretion**, and **reproduction**. Along each side of the insect abdomen can be found spiracles. **Spiracles** are holes in the cuticle that connect to the insect breathing tubes called **tracheae**. While insect breathing tends to be passive, some insects, at

THE WINGS HAVE IT

Wings have made insects the most successful animals on earth. Yes, it's true that most birds have mastered the art, and one mammal, the bat, is considered an accomplished flyer. But insects, in terms of numbers that have taken up flying, are the masters of the air.

Insect wings are as diverse as insects themselves. In fact, many of the insect orders used to classify insects are based on wing characteristics. Most order names include the Greek "ptera" which means wing. The ancient Greek philosopher Aristotle even classified insects according to their wing structure.

The typical number of wings for insects is four. However, the flies have only two wings. They are classified as Diptera which literally means two wings. The back pair of wings in the Diptera are modified into knob-like structures called halteres. Halteres function as balancing organs when flies are in flight.

The butterflies and moths, Lepidoptera, are known as the scale wings, because their wings are covered with scales. These scales provide the beautiful color patterns that make butterflies among the most recognizable insects.

Some insects like the cicadas, dragonflies, bees and wasps have four transparent wings. In the case of the cicada, the transparent wing allows the surface under the insect to show providing camouflage from potential predators.

Beetles have forewings that are hardened structures called elytra. These structures cover and protect the membraneous hind wings of these insects. It is the elytra that provide the look of "beetleness" that most of us readily recognize. It is that shape that has resulted in a well-known vehicle called the "beetle."

Some insects, the true bugs, like squash bugs, boxelder bugs and stink bugs have forewings that are part leathery and part membranous. When these bugs fold their wings, the membranous part is overlapped to cover the abdomen of the insect. Bugs are either right-winged or left-winged depending on which wing is placed on top of the other when they fold them.

Not all insects have wings. Fleas, lice and bedbugs have no wings at all. The lack of wings really hasn't hindered the ability of these little creatures to get around. They have taken to hitch-hiking as a preferred mode of transportation. However, there are far more insects using wings than hoofing it.

Flight is indeed the preferred means of transportation in most adult insects. "Have wings, will fly" seems to be the motto of the insect world.

INSECT LEGS ARE NOT JUST FOR WALKING

One of the defining characteristics of adult insects is that they have six legs. In some historical scientific circles, insects actually were called hexapods, based on leg number.

Some immature insects don't follow the six-leg scheme of adult insects.

Some immatures, like fly maggots, don't have any legs. Others, like caterpillars, have more than three pairs of legs. Technically, those caterpillars still have six legs. The things that look like legs are called prolegs, but they function in a similar fashion to legs.

Insect legs come in many sizes and shapes. You can learn a lot about an insect by looking at the way the legs are shaped. For instance, many insect legs are very much like human legs in structure. Both insects and humans have a leg part called a femur. We call it our thigh bone. Below that, in both insect and human legs, is a tibia. We call that our shin bone. The bones making up our feet are called tarsae. The same is true of insects. Our legs and the similar insect legs function to run or walk. Insects like cockroaches, ladybugs and fireflies have legs that function to walk or run.

Legs in insects exhibit many modifications from the basic walking shape. For instance, the front legs of camel crickets look like miniature shovels. The insect uses the front legs to dig its burrow. Dung beetles have front legs that are similar to those of the cave cricket. The dung beetles use forelegs to dig, but also to fashion balls of mammal dung. The dung, or manure, is buried in the ground and is the food for immature dung beetles.

Some insects, like crickets, grasshoppers, katydids and fleas, have hind legs modified for jumping. All of these insects are well-known for their ability to jump when attempting to escape from danger.

The front legs on some insects are designed for grasping. The praying mantis has legs like that. Mantids use their forelegs to catch and hold the insects that become their food. The same is true of other insects that make meals of their cousins. Water striders use their front legs for grasping, as do the predators known as assassin bugs.

Dragonflies and damselflies have legs that function mainly for perching on plants. But these insects use their legs to form a basket, which is used to trap small flying insects when the dragonfly overtakes them in flight.

Insects that live in water have legs shaped as paddles for swimming. Lice have legs that are designed to grasp the hair of their hosts. In fact, many insect legs have devices that allow them to cling to things—claws for hooking into something soft and sticky pads for hard, slick surfaces like glass. Many insects have both claws and sticky pads so they are equipped to walk on or hang on almost any surface.

Insects also use their legs for functions other than locomotion and grasping. Several grasshoppers, crickets and katydids have legs with combs that are used with other legs or wings to produce sound.

Some butterflies and flies have taste buds in their feet. This allows them to taste any item on which they are standing. For butterflies, they use the information to find a host plant for egg laying. For flies, the information tells them if the item they are standing on is good to eat. If so, they extend their mouthparts and begin to feed.

Other modifications of insect legs are to help camouflage the insect. Some insects that feed on plants have legs that resemble leaves to help them blend into their environment.

Another unusual modification of an insect's leg is found in the honey bee. Honey bees have a pollen basket on their back leg. They collect the pollen from their hair with a comb on a leg and place it in the basket for transport to the hive. So honey bees observed with yellow blobs on their legs are pollen collectors with their pollen baskets filled.

It is apparent that insects use their legs for many different functions other than just locomotion. They have to; after all, they don't have a pair of legs modified into arms as we humans do!

times, telescope the abdomen to force air into and out of the spiracles.

The end of the abdomen (which is also the end of the insect) is the location of the end of the digestive system, the **anus. External reproductive organs** are also found at the end of the abdomen in most insects (dragonflies are an exception). The most conspicuous reproductive organs are the ovipositors of the females. Some ovipositors that are hard not to notice are those which are modified into stingers such as in female bees and wasps. Some insects also have antenna-like sensory organs called **cerci** on their abdomens.

Insects, compared to humans, are said to be inside out. That's because their skeleton is on the outside instead of the inside like ours. The insect **exoskeleton** thus serves not only for support but is also the outer covering of the insect body. The exoskeleton of the insect is composed of several layers. The outer layer (called the cuticula) contains pigment and a nitrogenous polysaccharide called **chitin**. Chitin is insoluble in water and along with a wax layer provides the insect with protection against water loss. Chitin is very similar to another complex organic compound found in plants—cellulose.

The outer layer rests on a cellular layer called the basement membrane. When an insect sheds its exoskeleton, which it must do to grow, the basement membrane digests the lower portion of the cuticula and forms a new cuticle under the outer portion of the old skin. That old skin is then shed. The cast skin is called the **exuviae.**

The process of an insect (or other arthropod) shedding its skin is known as **molting.** The time following emergence from its old skin is a dangerous time for an insect. It takes a while for the new cuticle to harden, a process called **tanning**. During this time the skin is soft and the insect is a target for predators.

The internal systems of the insect, as in the human, can be described in terms of five primary functions. However the system in the insects is somewhat different than in humans.

Nutrient procurement and processing. Both insects and humans have a mouth and esophagus that leads to a stomach. Both have salivary glands

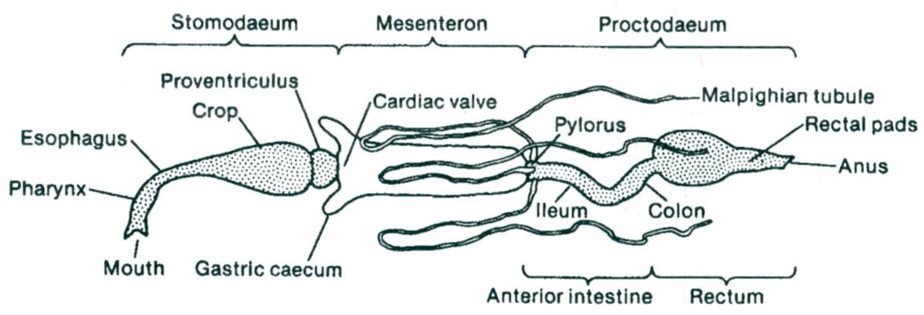

FIGURE 4.3.

that begin the digestive process in the mouth. In an insect, before the food gets to the stomach it enters first the **crop** and then the **proventriculus** (or gizzard), two organs insects have in common with birds. As food enters the insect stomach it is encased in a sac called the **peritrophic membrane**. From the stomach in both humans and insects food enters the intestines and then goes to the rectum and leaves through the anus. In insects the remainder of the digestive process is called frass—a fancy way to say insect manure!

Internal transport. Both insects and humans have an internal transport fluid. Both fluids are pumped by a heart but in the human the fluid is contained in a closed system (our veins and arteries). In insects the system is an open one. The fluid in humans is called blood and it carries oxygen from the lungs and nutrients to and waste products away from the cells. The internal transport fluid in insects is called **hemolymph** and it flows from the tube-like heart and bathes the tissues before it returns to the opening of the heart. The insect system is a bit like a pump submerged in a sac of liquid—the sac being the insect and the liquid the hemolymph.

Regulation of body fluids. Insects and humans both have fluid filtering devices. We humans have kidneys. Blood flows through the kidneys where the nitrogenous wastes are removed and excreted

Antennae Aren't Just for Space Aliens

Some people respond to insects as if they were creatures from another world. Maybe it's because they look otherworldly. The jointed legs, exoskeleton, multifaceted eyes and the antennae do give insects a far-out appearance.

To be sure, the human animal has eyes, legs and a skeleton, although somewhat different than those of insects. But when it comes to antennae, we humans have been shut out—dealt a big zero by Mother Nature.

We generally have less than an outpouring of enthusiasm for nature's creatures with antennae. That includes the insects. Maybe it's because insects wave their antennae in our faces from time to time. Or maybe it's because we're jealous that we don't have similar structures.

Actually we should be jealous. Antennae are marvelous structures that allow insects to gather information about their environment. Hairs on the antennae of some insects pick up vibrations in the air and allow the insect to hear. This is the way some male midges and mosquitoes detect their mates. The sound of the vibrating wings of the female are detected by the antennae of the male.

Both taste and smell are associated with antennae in some insects. The chemical clues detected include alarm chemicals produced by other insects of the same species. That is the reason ants very quickly come to the aid of a nest mate that has been injured or killed. Chemicals produced by female moths sometimes can be detected 5 miles downwind by a male of the same species—all because the chemical is picked up by his antennae.

In addition, insects are able to identify suitable plants on which to lay their eggs because of the odor of the plant. Many times, the plant odor is detected by sensors on the antennae. It is also the way flies find dead animals.

Some parasitic wasps actually use their antennae as a dog uses its nose, to trail another species. These wasps will tap their antennae on a surface as they attempt to follow a specific insect. When they find it, their goal is a sinister one. They will try to lay an egg on or in the creature—an egg that will hatch into a larva, which will feed on the parasitized insect.

There are many kinds of insect antennae, and all of them have names. For instance, many grasshoppers, crickets and cockroaches have long antennae with segments that are about the same size from base to tip. These threadlike structures are called filiform and are great for waving around.

Some antennae, such as those of butterflies, have knobs on the end and are called capitate. The antennae of moths don't have knobs on the end. Instead, they generally are fuzzy, a condition termed plumose.

Some insects seem to have flags at the end of their antennae, which are said to be lamellate. Some are bent in the middle and are called geniculate. Such is the case with ant antennae. That is good to know because that is one way to tell an ant from a termite. Antennae of termites aren't geniculate; they are more threadlike.

Whatever they are called, those insect antennae are busy picking up signals—hopefully not from outer space!

Cold Blooded Insects

When you hear the term "cold-blooded," what comes to your mind? Someone who gets cold easily? Or someone who is unfeeling, uncaring or ruthless? Of course, some of you more hip readers may think of Paula Abdul's song "Cold-Hearted Snake." But most of you probably don't think of insects.

Biologically, humans are not cold-blooded creatures. We maintain our body temperature near 98.6 degrees Fahrenheit winter or summer. But not so for snakes, salamanders, frogs, earthworms and insects. These animals are part of nature's "cold-blooded" approach to life. Their body temperatures are near that of their environment.

Such a lifestyle does have its disadvantages, but insects manage to cope. Everything happens more quickly in insects as the temperature increases. They move faster, eat faster and even grow faster. This means that insects complete their life cycles in less time when the temperature is warm than when it is cool.

When temperatures are too low for activity, some insects rely on the sun for help. They sunbathe, soaking up a few rays to increase their body temperature. These insect sunbathers are called baskers. Like human sunbathers, some insect baskers bathe with their backs in the sun. These are called dorsal baskers. Others prefer the rays on their sides—the lateral baskers.

Some other insects increase their body temperature by vibrating their wings. Moths and bees can frequently be seen with their wings quivering in the cool morning air or in autumn. Such activity is akin to warming up an aircraft engine prior to flight. But with the insect, as soon as the body temperature is high enough, it's off into the wild blue yonder.

Honey bees have a sustained flight temperature threshold of about 54 degrees Fahrenheit. When the air temperature drops below 54 degrees, honey bees may not be able to fly. It is for that reason that some honey bees will be found away from the hive during the cool evening hours of spring or fall. These bees leave the hive when temperatures are suitable for flight, but are unable to return when temperatures drop below threshold. Such is the risk that a cold-blooded insect takes when flying near flight threshold temperatures.

Of course in human terms, if the insect is really cold-blooded, it won't matter anyway.

A Tale of Blood and Guts of the Insect Type

Some time back, I received a letter with a request to identify insects that a homeowner had discovered in his house. Were they ants? Termites? Or something even worse? The insects were enclosed in a small plastic bag. The envelope—and its contents—had been processed through an automated stamp-canceling machine. And, in this case, more than the stamp was cancelled. The insects were smashed to smithereens.

Now smashing insects is not something new to humans. We smack insects every chance that we get. Fly swatters were invented for just that purpose. But we will use our hands, newspapers and even our hats as insect-whacking devices when necessary.

Whenever we are successful in smacking an insect what remains doesn't bear much resemblance to the intact creature. To be sure, you can sometimes find an identifiable wing, a leg or two, and maybe even an antenna, but the most obvious remnant is a greasy, gooey little spot. We're talking insect blood and guts here.

Most people have some notion about how the outside of an insect looks. You know, it has three major body sections. There is a head in the front with a mouth, eyes and antennae. A middle section, with wings and legs attached, called the thorax is next. Then comes the abdomen, which literally brings up the rear. But most people don't think much about insect guts. At least, not until they look at their hand after terminating a fly or when they struggle to clean the car windshield after a warm summer night's drive.

The blood and guts of insects are every bit as important to them as our innards are to us. However, there are some differences. Both insects and humans have important fluids that flow through the body. We call ours blood. In insects, that fluid is called hemolymph.

Human blood flows through a series of tubes. On the way to the heart, those tubes are veins; away from the heart and lungs, the tubes are arteries. Insects don't have a series of tubes to transmit such fluid. The insect hemolymph just sloshes around in a bag—a bag that we call the insect body. Scientifically, the human system is closed. Insects, therefore, have an open system.

Closed and open is not the only difference in the internal fluid systems of insects and humans. In mammals, including humans, blood carries oxygen to the cells and carbon dioxide away. The exchange occurs in the lungs. It is a red substance called hemoglobin that is the carrier, which is why our blood is red.

The insect blood does not carry oxygen. It lacks hemoglobin and is not red, except in a very few insects. For example, aquatic larvae of a group of flies called chironomids are known as bloodworms because they have hemoglobin in their blood. This is also true of the larva of a bot fly and a bug called a backswimmer. But the hemolymph of most insects is either colorless or yellow in color.

Another major component of insect innards is the digestive tract. It would normally contain food in various states of digestion. The inside of an insect would also include what is known as fat body. Insects, like humans, store energy in the form of fat. In insects, the fat is stored in specialized cells that are located throughout the inside of the insect.

So the next time you see a smashed insect don't just think greasy, grimy insect guts. Be scientific and think hemolymph, undigested gut contents and fat body cells! And look closely, because there ought to be some wings and legs thrown in there somewhere.

in urine through what is called the urinary system. The insect kidney is called the **Malpighian tubules**. Named after an old time Italian scientist the Malpighian tubules remove the nitrogenous wastes, but since the insect does not have a separate urinary system the waste is excreted through the intestine (same as in birds).

Gas exchange. Gas exchange in fine folks everywhere is achieved through the lungs (that's why you shouldn't smoke). In the lungs the blood is oxygenated and it carries the oxygen to the cells. Insects lack lungs. Oxygen is supplied to cells by **diffusion through the tracheae**. This requires that all cells be within diffusion distance of air. This means that insects have to be small for the system to work.

Nervous system. In both insects and humans the basis for the nervous system is the nerve cell and synapse. The chemical and physical nature of both systems is identical. That is why insecticides that attack the nerves work equally well on humans and insects.

Both insects and humans have a brain and a central nerve cord. The major difference in the nervous system of the two organisms is where the central nerve cord is located in the body. In humans the central nerve cord (the spinal cord) is located on our back or dorsal side. In insects the central nerve cord is located on the ventral side. Relative to the heart, the position is reversed in humans and insects. Compared to humans the insect is built upside down.

Insect Structure and Function

1. Name the three major parts of the insect body. What is the primary function(s) of these body parts?
2. What does the Greek word pterum mean? Where in the study of entomology do you commonly find the use of this Greek root?
3. What name is used for the breathing holes of insects?
4. The cerci that are found on the south end of an insect headed north are analagous (correspond in some respects) to what other insect structures?
5. Are insect wings modified legs?
6. What are apterous insects? Name some.

CHAPTER 5

Insect Movement

When it comes to "gittin' along" insects rank right up there with "little dogies!" Insects walk, crawl, fly and some even hitch a ride, but they manage to get there just the same. Insect movement can be classified into three general categories: seeking something good, avoiding something bad, or moving to new habitat. How this is done varies from insect to insect and among developmental stages of the same insect.

In insects, as in humans, crawling precedes walking. Most immature insects do an insect version of the crawl. Maggots, larvae of flies, are legless so they have to wiggle and writhe through their habitat. Bee and wasp larvae are also legless but they normally don't have to go anywhere since their sisters bring food to them. Many grubs have legs, but some don't use them effectively because of the shape of their bodies. Indeed one soil-dwelling grub uses its legs to sort of swim through the soil using a backstroke.

Caterpillars are the most interesting of the immature insect walkers. Caterpillars have six legs—the same number as the adult insect—on body segments very near the head. However the caterpillar has some extra leg-like structures further back. These of course can't be legs since insects can only possess six (otherwise they would be spiders, crayfish, or centipedes). So entomologists have overcome this obvious contradiction by calling the extra legs prolegs! The last two prolegs on caterpillars are used for clasping—hanging onto sticks and the like. In general, caterpillars walk by enlarging and contracting body segments and sort of pushing segments along in the general direction of the head.

One of the modifications of this push scheme to caterpillar walking is with a group of Lepidoptera larvae called loopers. Loopers are so called because they grasp the terra firma with their true legs (they have six of those, you know) and then loop their body forward. This brings their terminal pair of prolegs alongside the real things. At this point the caterpillar hangs on with the prolegs and propels the rest of the beast forward. This produces a looping motion, hence the name. Some loopers are called measuring worms because the worms appear to be taking a measurement as they loop forward.

Walking, in adult insects, requires coordination of six legs. The common method used by insects to keep from getting tangled up is the tripod method. In this case, the insect uses the middle leg on one side with the outside legs of the other to move forward using the single leg as a pivot for the tripod. However if the insect needs to move quickly it sometimes uses a system where all three legs on one side move forward at the same time. In the military this would be something like: your left, left, left, your right, right, right!

Flight is the real accomplishment of insects. This is powered by direct muscle contraction in dragonflies, grasshoppers and butterflies, in which case the maximum wingbeat frequency is about 30 beats/second. Other insect groups use indirect flight muscles where the muscles are used to contract the sides of the thorax in a spring-like fashion. Such a system can produce frequencies of up to 1,000 beats/second in some of the flies. How fast can a fly fly? Some horse flies can go 9 mph. Bumble bees can do better: 11.2 mph. But the hawk moth has been clocked doing 33.7 mph—hopefully not in a school zone.

Other insects hitch a ride, especially when moving to a new habitat. This is always true of fleas and lice. They just ride on the host. Some insects catch a ride on the wind. Winds bring more than rain when they come into an area. It is not a free ride, since most insects have to flap their wings to keep aloft in a moving air mass. When they get tired or stop flapping for whatever reason they drop out like rain. Hopefully it is to find a great place to live.

Have Wings, Will Fly

Birds do it. But so do the bees. In fact, most insects are real experts at it. They've been cruising the airways of planet Earth for 350 million years or so.

Insects have so mastered the art of flight that it is one of the reasons for their success. Indeed, in honor of their winged proficiency, we call some of them "flies."

As landlubbers, we humans stand in awe of the aerial gymnastics performed by insects. Scientists have studied insect flight and even counted the number of beats of an insect wing. A swallowtail butterfly, for instance, will flap its wings about nine times a second. A honey bee has a wingbeat frequency of 250 beats per second; the old house fly, about 190; and the culex mosquito, a little over 300. It is the frequency of the wingbeat that produces the mosquito's all-too-familiar humming sound.

House flies cruise along at nearly 5 miles per hour. Some sphinx moths are the fastest insect fliers, reaching a top speed of over 33 miles per hour. The honey bee can fly about 8 miles per hour, and the bumble bee can buzz along at 11 miles per hour, a respectable number considering that a scientist once calculated that a bumble bee couldn't fly at all.

It seems that the surface area of the bumble bee wing was too small to carry that fuzzy, yellow- and black-striped body aloft. Of course, folks who have been dive-bombed by a bee intent on causing bodily harm have experienced the error of that calculation firsthand!

Insects use the power of flight for many purposes, including finding mates, avoiding enemies and moving to a new territory. Some insects travel many miles on the wing. The champion traveler in this regard is the monarch butterfly. This insect travels each year from its breeding grounds in Canada to overwintering sites in the mountains of Mexico—a trip of over 2,000 miles.

Of the teeming billions of insects, not all use flight as a preferred form of locomotion. Some prefer to hoof it, as this old poem tells:

The June bug hath a gaudy wing,
The lightning bugs a flame,
The bed bug hath no wing at all,
But he gets there just the same!

Out of Sight, Out of Mind

Most of us recognize that most adult insects have wings. The ability to fly is, after all, one of the secrets to the success of the insect world. Yet some insects are very successful without wings. Fleas and lice come to mind.

Scientists believe that fleas and lice at one time had wings, but have lost them because these appendages weren't needed. Fleas and lice live on animals that are densely covered with hair or feathers. Insects can't fly in such an environment, and wings just made it more difficult to walk.

A few insects, on the other hand, never had wings. They are known as Apterygota—without wings. There are three orders of insects classified as Apterygota. Included in these orders are silverfish, firebrats and collembolans.

Silverfish and firebrats are domestic species that inhabit our homes. Silverfish are gray, firebrats are tan or brown, and both have long antennae and three tail-like appendages. Silverfish prefer a moist area, so we encounter them in our bathrooms and under kitchen sinks. The firebrats, as the name suggests, like warm places, so they are found around fireplaces and furnaces. Both species are active and run rapidly.

The collembolans, on the other hand, go largely unnoticed by the human population, even though they may be the most common of all insects. They are unnoticed because they are small and live in leaf litter, damp soil or rotting logs. Most of us don't spend much time looking in these places, so we don't see collembolans.

Collembola are commonly known as springtails. The name is based on the behavior of the little creature. Most species of Collembola have a forked structure on their abdomen that is folded forward and latched down. The structure is like a spring, and when released it projects the little creature forward—more like catapults it forward. Some springtails that are less than one-tenth of an inch in length may end up over 4 inches away following the release of the spring.

While most of us have never knowingly seen collembolans, there are hundreds of them in almost every handful of leaf litter. Sometimes they come out of hiding. Elliott Maynard, in his book "The Collembola of New York State," recounts reports of high numbers of these insects. In North Carolina, the numbers of one species were so high they formed blankets of insects inches thick. And in another instance, collembolans were so thick that locomotive engines in Switzerland could not move because smashed insects made the tracks too slick for traction.

Mostly, though, collembolans just quietly go about their job as one of nature's recycling crew, and hardly anyone notices.

Insect Movement

1. What name is used for the "extra" legs of caterpillars?
2. What geometric shape is the basis for adult insect walking?
3. What insect is reported to be the fastest flyer among all insects?
4. What insect travels further under its own power than any other?
5. Why was it said that a bumble bee should not be able to fly?

CHAPTER 6

Insect Growth

Metamorphosis

The life cycles of insects involve a process called metamorphosis, which is a change from one life form to another. Insects are not the only animals that undergo metamorphosis. Frogs, toads and salamanders change from water-breathing tadpoles into their adult forms that breathe air. The process of metamorphosis is widespread among insects. The process is said to benefit the insect because the juvenile does not compete with the adult for the same food resource. For example, the immature form of a butterfly, a caterpillar, feeds on leaves, the adult on nectar. The larvae of mosquitoes feed on microscopic plant and animal life in water, the adult on nectar and blood.

Insect life cycles can generally be divided into two groups. Those with a complete life cycle go through four stages of development—egg, larva, pupa and adult (Figure 6.1). In most instances, unless you have studied the larval stage of an insect, it would be impossible to know what kind of an adult insect it would become. Because of this mystery the ancient Latin word for this stage is larva (meaning ghost or mask)—the stage covered up or masked what it was going to be, what it would look like in the adult stage.

The table on the following page presents a summary of the stages included in the life cycle of insects with complete metamorphosis.

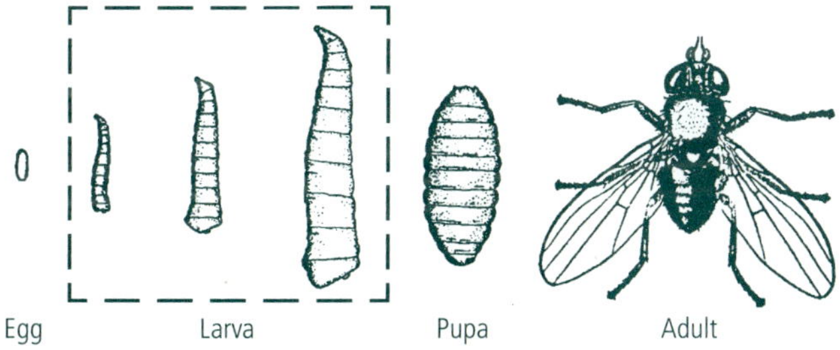

FIGURE 6.1. *Complete Metamorphosis.*

Insects that do not go through all four stages during development are said to exhibit incomplete metamorphosis (get it?—not complete!) because they do not have a pupal stage (Figure 6.2). In these insects there are three stages—egg, nymph and adult. The nymph (from the Greek, meaning bride) looks much like the adult but is not sexually mature and has no wings or only partially developed wings. In this case, such as with many plant bugs, both the immature and adult feed on the same food.

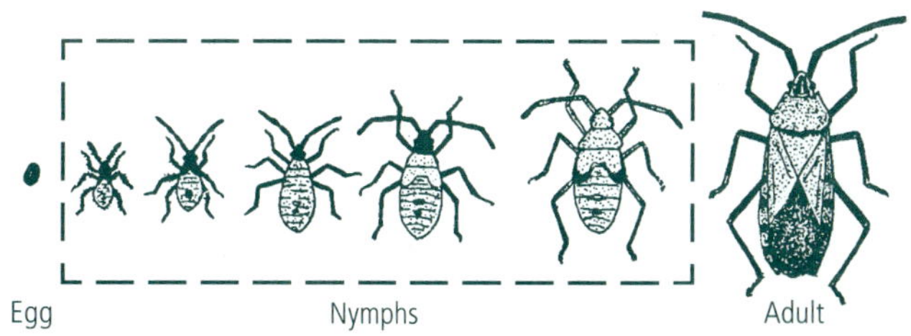

FIGURE 6.2. *Incomplete Metamorphosis.*

Presto Change-o!

Long before Dr. Jekyll transformed into Mr. Hyde, some insects had already mastered the ability to change their personalities.

These insects use a unique biological process called metamorphosis to change from one life form to another. By this process, a "Mr. Hyde," or a larva that can damage plants by its feeding, can become a Dr. Jekyll, or a usually harmless and sometimes beautiful adult insect.

Franz Kafka addressed the concept of metamorphosis in his work of fiction by the same name. A popular movie of more recent times, "Cocoon," also drew parts of its theme from the process of metamorphosis. Even yuppies have noted the parallel between a quiet time in their own hurried lives and the apparent quiet of an insect undergoing pupa-

	Stage			
	Egg	**Larva**	**Pupa**	**Adult**
Function		Feeding	Transformation	Reproduction
Characteristics	Inactive	Active	Wiggle	Active
Growth	Yes	Yes	No	No
Names	Nit	Grub Maggot Caterpillar	Chrysalis Puparium Cocoon	Imago

tion. In yuppie lingo, a lazy day of lounging at home is known as "cocooning."

Changes in the insect during metamorphosis are miraculous. The tissues and organs of the larva are broken down and reassembled. The result is to produce a winged, adult insect, completely different from the wingless eating machine of its youth.

The insect stage that harbors these changes is the pupa, from the Latin word for doll. Apparently, ancient people saw in that insect form a likeness to the inanimate toy for a child. The same Latin root is used to form the word "puppet."

Insects protect their pupae in a variety of ways. Some, like the common house fly, are thrifty, using their last larval skin to form a protective cover for the pupae. Other insects, like that pest of lawns—the white grub—industriously form a nice chamber in the soil at pupation time. Still others follow the Boy Scouts' lead and make a tent-like structure by tying the edges of leaves together.

Some insects spare no expense or trouble to make their cocoons. Probably the most famous of this group is the silkworm, which spins a cocoon of silk for pupation. Centuries ago, it was discovered that the silken thread used by silkworm larvae to construct pupation chambers could be unwound and used for human purposes. Since that time, uncounted silkworms have spun up cocoons with the biological intent of becoming an adult. However, humans have interceded to claim the silk for their own uses.

And then there are those insects that choose the no-frills method. Most butterflies—unlike their close relatives the moths—have pupae that are naked. As larvae, these streakers among the insect pupae normally attach their tail end to a leaf or twig before pupating. Then while the butterfly pupa, called a chrysalis, hangs upside down, Mother Nature gives the creature a makeover.

Of course, this is absolutely necessary. How else would an ugly caterpillar turn into a beautiful butterfly?

Autumn Wanderers

Autumn days are made for wandering. The air, so recently laden with the heat and humidity of the summer, acquires a refreshing crispness. The coolness of fall is a blessing and a curse. It relieves us of the summer swelter, but reminds us of the cold to come.

"As you go a-wandering," you might even be moved to sing. However, if a "Val-de-ri, val-de-ra" fails to escape your vocal chords in an oral celebration of the season, you are not alone. Many of nature's finest songsters—insects and birds—grow silent with the onset of fall.

The fall is a time when birds and insects prepare for winter. Some birds follow the sun and wander far from their summer homes. One North American insect, the monarch butterfly, heads south to Mexico. Most insects just prepare to wait the winter out.

For some caterpillars, winter preparation means looking for a suitable place to face the cold. The insects go "a-wandering" and are appropriately called wanderers. Many wandering caterpillars, but not all, go unnoticed by humans. Caterpillars of the giant silkworm moths are large, showy insects. Since they primarily feed on trees, they are not noticed by casual observers during the summer. As fall approaches, they leave their food plants and seek a sheltered site to spin a cocoon. This brings them down to our level, where they are found crawling across sidewalks, on porches and into garages.

Another group of highly noticeable insect wanderers includes the small, fuzzy caterpillars. These

Strip Teasing—Insect Style

Insects and their arthropod relatives—shrimp, crabs, and lobsters—possess an exoskeleton, a combination skin and skeleton which protects them against crushing, scratching, invasion by microorganisms, and water loss. A hard, durable exoskeleton is a great protective device, but it is a real problem as the insect grows—it won't stretch.

Knights of olde who covered themselves with suits of armor had a similar problem at times. Indeed, the knight who forgot to mind his waistline discovered the need for a new suit. Insects face a similar problem several times during their life cycles, and like the ancient knights, they acquire new suits of armor.

Before an insect grows a new exoskeleton, it sheds the old one, a process called molting. Most people are familiar with the shed skins of insects. For instance, what youngster hasn't gathered the conspicuous shells of cicadas? When the immature cicada nymph emerges from its underground home, it immediately molts to the winged adult stage. The old exoskeleton can be found fastened to a tree, shrub or fence post. Aphids also molt, and sometimes the only evidence of these pest insects on a plant will be hundreds of white, ghost-like skins.

Insect molting has long fascinated scientists, who have worked hard to describe the chemical nature of the process. Even poets have noticed—and described—the magic of an insect shedding its skin. In Tennyson's "The Two Voices," we read about a dragonfly molting:

An inner impulse rent the veil
Of his old husk; from head to tail
Came out clear plates of sapphire mail.

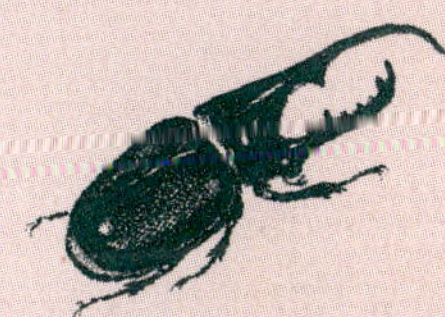

The ancient Greeks were aware of the molting process and had a word to describe it—ecdysis (eck DEE sis). It was from this word that H.L. Mencken coined his humorous term for a strip teaser—an ecdysiast!

The practical nature of an insect molting might not hold the same attraction to most people as the dance of a strip teaser. However, the result is the same; both the insect and the ecdysiast end up removing their outer clothing.

moths-to-be overwinter in some protected site. They seem to have difficulty finding just the right place to spend the winter, so they frequently wander about looking for another site. One group of caterpillars—the wooly bears with black and brown stripes—is famous for predicting the severity of winter. Their fame is probably enhanced by the fact that they wander on most sunny days late into the fall, long after much other insect activity has ceased.

A little wandering on a fall day is good for humans and insects. "Val-de-ri, val-de-ra!"

Metamorphosis

1. Why is metamorphosis said to be a benefit to insects?
2. What are the four life stages of insects that have complete metamorphosis?
3. What is the primary biological function of insect larvae?
4. What Greek word meaning bride is used to describe a form of immature insect?
5. The word "puppet" has the same Latin root as what insect stage?
6. Which group of Lepidoptera have naked pupae?
7. What two terms are used for the process in which an insect sheds its exoskeleton?

UNIT 3

The Types of Insects

CHAPTER 7

Velvet Wings (Lepidoptera)

Butterflies and moths are probably the world's favorite and most known insects. Children love to chase them, adults plant gardens to attract them, and artists often incorporate them into their work. Many bookstores have a variety of field guides and other publications dedicated to these delightful creatures. Butterflies and moths belong to the Order Lepidoptera (from the Greek *lepido* meaning scale and *pteron* meaning wing). Presently, there are over 120,000 species described, with over 11,000 occurring in North America (excluding Mexico, which boasts 25,000 species). Many adult butterflies are bright in color, which is to their benefit because they are active during the day, or diurnal. Most adult moths are active during the night, or nocturnal, and are usually dull in color.

Adult Lepidoptera have two pair of wings that are covered with scales that act like shingles on a roof. These scales are most easily observed as "dust" on fingers after handling a specimen. When spread, their wing span can range from 5 mm, about the size of a grain of rice, to 270 mm, the height of this book! The mouthparts of adult butterflies and moths are modified into a long siphoning tube called a proboscis. The proboscis is coiled under the head in a spring-like fashion while at rest, and then extended into a "straw" to extract nectar from flowers. Adults have six well-defined legs, and their bodies are generally covered with hair and scales.

Lepidoptera go through complete metamorphosis. Their eggs hatch into tiny plant eating machines called caterpillars. Unlike the adult sucking proboscis, caterpillars have chewing mouthparts. These caterpillars eat and eat and eat, and eat and eat and eat, and eat and eat and eat, dramatically increasing in size. To help with holding on to their food source, caterpillars have prolegs on their abdomens. These prolegs are fleshy extensions that end in little hooks called crochets. Try pulling a caterpillar off your finger slowly to feel these crochets "holding on." When the caterpillar is finished with its eating, it progresses into its next life stage, the pupa. The pupal stage is where a caterpillar transforms into an adult. For moths, the pupa is often called a cocoon, and for butterflies, this stage is often called a chrysalis. While the adults feed, they help pollinate the world of flowering plants.

Butterflies and Springtime

The signs of spring: little frogs, called spring peepers, croaking at the top of their amphibian lungs. Redbirds singing from the highest branch of the tallest tree. Bursting buds on the maple tree. Daffodils, tulips and violets. And butterflies.

There's nothing like the sight of a butterfly to indicate springtime. For many centuries, people everywhere have noticed the first butterfly of the new season.

The term butterfly probably was derived from the common yellow cabbage butterfly because it is first seen in the early spring season. Early spring has been called the butter season because milk production tends to increase then. In Anglo-Saxon language, these insects were called "butter-fleage"—meaning butter fly. A German name for these insects is "Schmeterling"—from cream—suggesting a derivation similar to the English term.

In most temperate regions of the world, butterflies overwinter as mature larvae or pupae. So when spring rolls around, they are ready to pupate and emerge as butterflies.

Butterflies are well-known seasonal travelers. The monarch butterfly goes north in the spring and south each fall across North America. In South

America, travelers in the jungles can sometimes determine the points on the compass by observing the direction of the flight of butterflies.

Whether as newly emerged adults or as migrants headed to summer breeding grounds, the presence of butterflies is a sure sign of spring in North America.

Many folk sayings are associated with the first butterfly of the season. In Devonshire, England, people are admonished to "kill the first butterfly you see each year, or you will have bad luck all through the year." This may be a reflection on the pest nature of the cabbage butterfly; its larvae feed on and damage many plants of the cabbage family.

In some areas of the world, good luck is said to follow when a butterfly lands on a person. That saying was known to a congressman who, while in attendance at a Washington political rally in the late 1800s, noticed that a butterfly had alighted on his hat. The fellow stated that his luck was going to be good. However, another congressman of the opposition party commented that he wouldn't

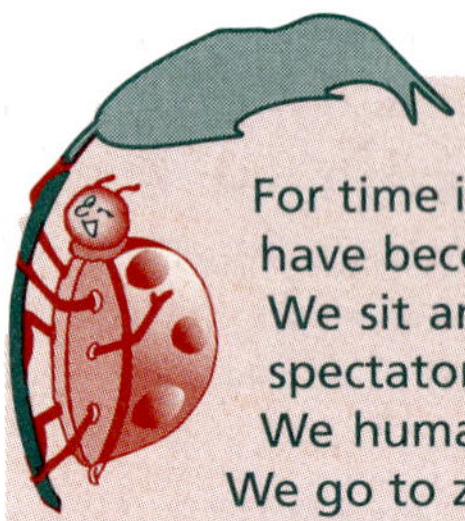

Butterfly Watching Is the Ultimate Spectator Sport

For time immemorial humans have enjoyed watching sporting events. So much so that spectator sports have become billion-dollar industries in modern times. We pay dearly to witness such contests.

We sit and watch baseball, basketball and football games. Soccer, cricket and tennis matches attract spectators galore. So do ice hockey and golf. Millions witness horse and auto racing.

We humans love to watch things. We watch fireworks displays, plays and even construction projects. We go to zoos and watch the animals.

Humans even make a sport out of watching each other. People watching at the mall is a favorite pastime for many of us.

Watching the opposite sex has always had a special allure for humans. A top musical hit for The Four Lads in the '50s was "Standing on the Corner." The song featured girl watching—"Standing on the corner watching all the girls go by!" In 1968, the one-hit-wonder the O'Kaysions released "Girl Watcher" that began with the line, "I'm a girl watcher, Watching girls go by." But we all know that boy watching is just as popular.

We also like to watch animals in nature. Bird watching is a very popular hobby—and so is butterfly watching.

Butterfly watching is becoming an increasingly popular sport. It has even entered the realm of a paying spectator sport. Butterfly houses—facilities that have free-flying butterflies—have become popular tourist attractions in several major cities around the world. For a fee, humans get to interact, up close and personal, with beautiful butterflies.

You don't have to travel to a commercial butterfly house to be a butterfly watcher. At least in the good old summertime, all you have to do is look in your backyard. For most of us, the summer brings the butterflies to our yards and parks—at least if those yards and parks have flowers.

Some people encourage the butterflies with plants that support the butterfly populations. Such butterfly gardens are becoming increasingly popular.

In butterfly watching, as in other sports, the more you know about the game the more enjoyable it is to watch. So it helps to be able to identify some of the more common butterflies.

Unlike participants in many sports, butterflies don't have uniforms with names and numbers on their backs. But their uniforms are distinctive, so you can identify butterflies by the way they look. This sometimes is aided by having a pair of close-focusing binoculars.

Some of the common butterflies that we see in the Midwest belong to one of the following groups. The swallowtails are our largest butterflies. Their long tails give them their name and make them easy to identify.

The whites and yellows are very common. Their names reflect their colors and they exhibit a low and quick flight. They can be found all summer long.

Angle wings and tortoise shells include the question mark and mourning cloak butterflies. These butterflies can frequently be seen feeding on rotting fruit or carrion. They are commonly seen resting on dirt roads.

The skippers are brownish butterflies that exhibit a quick, darting flight. They all have hooked antennae that make it easy to identify this group.

The blues, as their name suggests, are blue colored. They are fairly small butterflies. The blues and angle wings can frequently be seen drinking water around a mud puddle.

Most people recognize the monarch, or milkweed, butterfly. It is frequently observed on the southward migratory flights each September and October.

Butterfly watching is a fun spectator sport. And the price is right!

Butterflies Are Favorite Insects for Most People

Most people don't care much for insects. An exception is the insect that some poets have called a flying flower. Yes, almost everyone has a soft spot in his or her heart for butterflies.

If any insect group is perceived by humans to be warm and fuzzy, it is the butterflies. Butterflies really aren't warm, though. Like all insects, they are cold-blooded. But they are fuzzy! Butterflies, and their insect cousins the moths, have fuzzy wings.

Butterflies and moths are classified in the same insect order. The order name, Lepidoptera, reflects the fuzzy nature of their wings. It means "scale wing." There are roughly 10 times as many species of moths as butterflies, but most people are more familiar with the butterflies.

That is partially because butterflies are active during the day while moths are active at night. In addition, butterflies are mostly bright-colored. Moths tend to be somewhat drab. Consequently, people are more likely to notice butterflies than moths.

The color of butterfly wings is achieved in one of two ways. Some of the color is in pigments, much as it is in most other animals. In addition, color in some of the brighter butterflies—and a few other insects—is due to light reflection. In butterflies, small parallel ridges on the wing scales break up the light into component parts. Water droplets in the air do the same thing to light when a rainbow is produced.

The color of the beautiful morpho butterflies of the Amazon is produced in this way. Because the color is the result of light reflection, the color changes when these butterflies move their wings. Many morphos are a bright, metallic blue when the wings are held out from the body. When held vertically, however, the bright color disappears and the wings appear a brownish gray.

Morpho butterflies tend to slowly open and close their wings when at rest. Because the color changes as the wings move, the behavior works to confuse a potential predator, which might be considering a butterfly meal. The butterfly is playing a now-you-see me, now-you-don't game with the predator.

The same effect is achieved by butterflies that have bright colors on the outside of their wings but duller colors on the underside. When the wings are folded, such as when the butterfly is resting, the color pattern may allow the insect to blend into the environment. Such is the case with the angle-wing butterflies. These butterflies are medium-sized of a reddish-brown color with black spots. The undersurface of their wings is gray and brown, colors that imitate bark and dry leaves.

Butterflies are often named after the way their wings look. Most people would recognize that swallowtail butterflies are named after a projection, which looks like the tail of a bird known as a swallow, found on the hind wings.

The angle-wings have wing margins that are scalloped and indented as if they had been cut with scissors.

The whites and sulfurs are common butterflies, with names based on the color of their wings. There is a group of butterflies called the purples, and, you guessed it—they are purple in color. There is a banded purple, a spotted purple and a great purple hairstreak.

Peacock butterflies are dusky brown in color but have eyespots on the wings, similar to those found in the feathers of peacocks. Another group of brown-colored butterflies is the tortoise shells. Of course, their wing pattern reminds folks of the shell of a tortoise.

What is it that people like about butterflies? It might be the bright colors. Maybe it is their delicate nature. It might be the fact that they visit our gardens. But I suspect that people really like butterflies because—unlike some other insects—these flying flowers can't bite or sting!

be so sure. It was a cabbage butterfly that was on the hat. The insect, the observer noted, had recognized a cabbage head when it saw one!

Tenting on the Old Campground

For many folks, spring means the return of the camping season. It's the time to retrieve the musty-smelling tent from the basement or attic. But before most human campers make their first spring foray into the great out-of-doors, the insect tents are already in place.

The most conspicuous of the insect tentmakers is the eastern tent caterpillar. This insect, a hairy brown caterpillar that grows to about 2 inches in length, has been observed making tents in America's trees since 1646.

The insect passes the winter in the egg stage. In early spring, the eggs hatch. Young larvae band together and produce a large thick web, called a tent. The tent is normally located in the fork of a tree.

Larvae do not feed within the tent. They use the tent for protection from the weather, and possibly from predators. Hungry larvae leave the tent where they feed on the newly forming leaves of their host trees. Favorite trees are wild cherry, apple, peach, and plum, although other trees are sometimes attacked.

By midsummer, the larvae pupate and turn into yellowish brown moths. The moths mate and lay clusters of eggs which will hatch the next spring to start the cycle over again.

Some people, it seems, do not like the sight of a group of happy campers—at least of the insect variety—so they try to dispose of the little beasts. If the tree under attack is small, one approach is to prune the limb with the web and worms and destroy it. Several insecticides can also be used to kill the caterpillars. However, treating large trees with insecticide requires specialized equipment, so it may be necessary to hire a professional.

In general, the feeding of tent caterpillars will not cause permanent harm to a tree. Usually the tree will have time to leaf out following the feeding. However, defoliation by tent caterpillars in two or more consecutive years could stress the plant and warrant control.

So the damage caused is more to the aesthetics of the tree and the pride of the owner. However, some people take the approach that a few insect tents in a tree are an interesting biological event—especially if they can't be reached with the pruning shears.

Cutworms Everywhere

Gardeners hate them, and farmers live in fear of them. They are cutworms, larvae of moths that are named after their habit of cutting plants as they feed. There are many species of cutworms, and their numbers vary greatly from year to year.

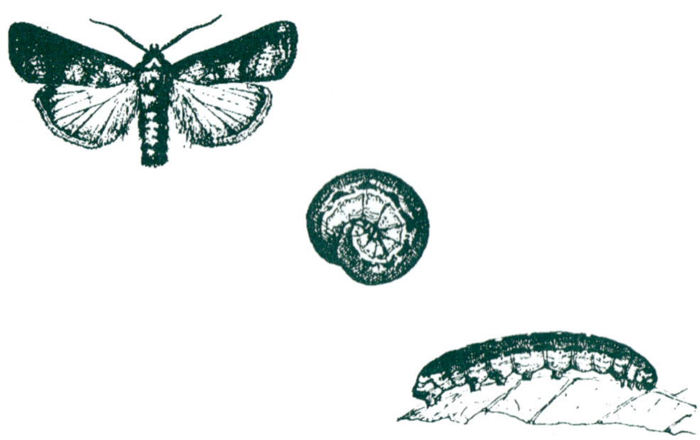

Some cutworms are described in terms of their lifestyles. There are climbing cutworms, solitary cutworms, army cutworms and subterranean cutworms. Some cutworm names are descriptive of their appearance, such as spotted, black, dingy, yellow-striped, white, glassy, or bronzed cutworms. But generally, cutworm moths are rather nondescript, drab-colored insects. The moths are normally grey and brown with few distinct markings.

Most cutworms spend the winter in the larval stage in temperate regions. In the spring, the larvae pupate and emerge as moths, which then lay eggs in areas of vegetation. There, newly hatched larvae begin to feed on the closest plants.

One of the most widespread cutworms is the black cutworm, which is found worldwide. In the heartland of the United States, the black cutworm is a seasonal invader. Moths migrate into the area each spring from areas of the Gulf Coast and lay eggs in early-season weeds or even crop debris.

When a new crop is planted, the hungry larvae begin to feed on those plants. These young larvae feed on foliage for a week or so, then their feeding habits change. Black cutworm larvae become light-sensitive and go underground. The insect, which is now termed a subterranean cutworm, emerges from its soil hideaway during the nighttime hours to feast upon tender stems of all kinds of plants.

If the target of cutworm feeding just happens to belong to a gardener, an irresistible force has met a self-proclaimed immovable object. The gardener cherishes each plant. The cutworm looks at all plants as mere food items and proceeds to do what it does best: eat! What's more, the cutworm adds insult to injury by cutting more plants than it really needs for a food supply. Much to the dismay of the gardener, this results in dead and damaged plants, so chemical warfare begins.

It's a timeless battle between cutworm and gardener, and one thing is for sure: gardeners will, at some time, get caught with their plants down!

Ugly Worms

True silkworm moths are not native to North America. However, the largest and most-attractive moths on this continent are called giant silk moths.

The giant silk moths vary in size with wing spans from 3.5 to 6.5 inches. These moths include the Cecropia, the Promethea and the Polyphemus, named after Greek mythology gods.

Adults of the giant silk moths do not feed. Their mouths are barely developed and not capable of sucking up nectar as is done by most moths. Their immature forms, however, make up for the lack of feeding on the part of their parents. Larvae of giant silk moths are voracious eaters, feeding on leaves of trees and shrubs. These large larvae are spectacular insects. Most are green and ornately armed with brightly-colored tubercles and spines. Some folks think these larvae are downright ugly!

During the summer months, the larvae are small and generally go unnoticed by the public. However, they grow quickly and as fall approaches, the small worms have turned into big worms. Some are 5 to 6 inches in length, and are quite visible as they wander around looking for a suitable site to spin a pupal case.

These "wanderers" are sometimes discovered crawling across sidewalks or up the sides of houses. Such encounters between humans and giant silk moth larvae elicit less than appreciative responses from most folks. Many people are inclined to describe these large, spine-bedecked larvae as gross, disgusting or worse.

"Ugly is as ugly does." In this case, being ugly discourages most predators from attempting to make a meal of the creatures. But the worm's appearance only fascinates little boys, who frequently carry them home for pets. Whereupon, long suffering mothers will come up with a makeshift cage for the new pet. The worm responds by spinning a cocoon. There, snug as can be in its silken case, the worm waits for spring. When spring comes, the ugly worm is miraculously transformed into a beautiful moth. Could it be that the "ugly duckling" was really the larva of a beautiful, large moth?

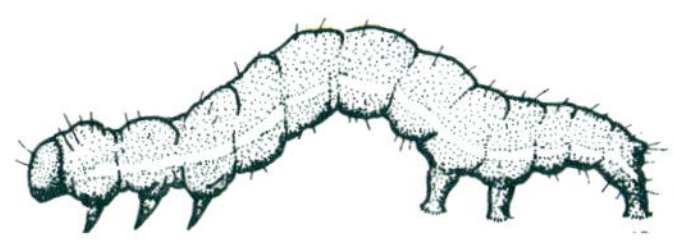

Caterpillars

Spring is the time of year when mother nature evicts old man winter from the seasonal driver's seat and hands the reigns to spring. It's a season of renewal—birds sing, flowers bloom, and butterflies and moths lay eggs.

The Death's-Head Moth

Remember the old saying "if looks could kill, you'd be dead?" Well, the death's-head moth has the look—the pattern of a human skull is well-defined on its back.

But don't be afraid; this insect is not some dastardly devious member of the insect world. It can't bite or sting, nor does it pose a sinister threat to humans. Unlike some moths, the death's-head moth isn't interested in destroying our possessions.

This large moth has a wingspan of 4 inches and is found throughout Europe. Technically, the death's-head is one of a family of moths known as sphinx or hawk moths. They are swift fliers and are found around flowers in the evening. Because of this behavior, they are sometimes called hummingbird moths.

Larvae of these moths are known as hornworms because they have a conspicuous horn protruding from their posterior. The familiar green worm that feasts from time to time upon tomatoes is a hornworm, the tomato hornworm to be exact.

Because of its name, the death's-head hawk moth is a well-known insect. Hood used the insect in his poem "Haunted House." More recently this insect has been featured in the movie *Silence of the Lambs*. In the movie's advertisement, the death's-head moth is shown across the mouth of the film's star, Jodi Foster. The insect plays an important role in the movie. In fact, it lives up to its name although through no fault of its own. The movie makers, however, while acknowledging the foreign nature of the real death's-head moth, appeared to have used its entomological cousin, a tobacco or tomato hornworm as a stand-in for the real thing. No matter; the real thing still gets the bum rap.

So how does the death's-head moth live? In its immature stage, it feeds on potatoes, and in its adult stage, it hangs around honey bee hives, perhaps hoping for a little sip of honey.

Unlike other moths, the death's-head squeaks like a mouse. A hired killer this insect is not. Its reputation is due entirely to the clothes it wears.

These butterfly and moth eggs will hatch into creatures called caterpillars, which have a very important biological function—eating. Their most common food is the green tissue of plants. For this reason springtime, when all the leaves and trees are green, is a great time to be a hungry caterpillar.

The word "caterpillar" is based on the Latin *catta pilosa* which means hairy cat. The exact origin of the term is unknown; however, hairy does apply to some caterpillars. Immatures of moths are generally covered with hair and include such familiar insects as the wooly bear. The majority of butterfly larvae are naked—without hair—but are still called caterpillars. Most caterpillars move using a creeping motion; therefore, ancient Romans may have related it to the stalking motion of a cat.

Caterpillars, of the insect type, are familiar creatures. Some folks find them chowing down on their favorite plants and become hostile to the little creatures. To other folks, a caterpillar is something to admire, or in the case of little boys, something to collect and take home in the pocket of a shirt.

Such admiration is expressed by Christina Rossetti in her poem "The Caterpillar."

Brown and furry
Caterpillar in a hurry
Take your walk
To the shady leaf or stalk
Or what not,
Which may be the chosen spot.
No toad spy you,
Hovering bird of prey pass by you;
Spin and die,
To live again a butterfly.

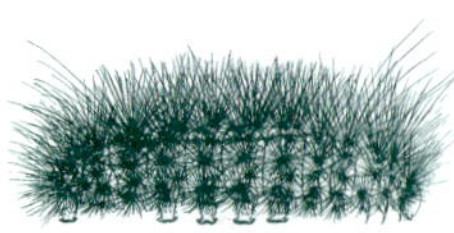

To most people the caterpillar is of the insect variety, but to others it is a yellow crawler tractor manufactured by the Caterpillar Corporation. Why such a name for a tractor? It seems that the early crawler tractors brought to mind the insect version, thus the name of the company and its products. Whether it's a tractor or an insect larva, the caterpillar can frequently be seen working in the green fields of summer.

Corn Borer Murder

If murdering a corn borer moth were a crime, all nighttime motorists would be in jail. Those cream-colored moths fluttering in snowstorm-like density in the glow of headlights are corn borers. They are the victims that end up on radiators, in grills, or as unsightly grease spots on windshields and headlights.

In their immature or larval stage, corn borers are major pests of corn. As worms, they bore in the corn stalk—that's the basis for their name. Such activity weakens the plants and can result in yield reduction. Those borers become moths, and the moths in turn lay eggs that hatch into borers; so the cycle goes.

In most years, the borers are still chomping on corn plants during late summer and fall in preparation for winter. However, an unusually warm summer accelerates development of many insects, including corn borers. Consequently, moths are still flying in August when they normally do not emerge until the following June. An early emergence coupled with high numbers creates a late-season "storm of moths."

Corn borer moths spend their days lounging around in grassy areas such as roadsides and creekbanks. When the sun goes down, these moths take to flight, seeking suitable plants on which to lay eggs.

Corn borer moths, like many other night-flying insects, are attracted to light. Consequently, moths are frequently found congregating around porch lights and windows far from their home cornfields.

However, drivers traveling through areas of high corn density are most likely to encounter the fluttering masses of moths. These insects are going about their "mothly" business when the auto invades their airspace. The stream of air flowing over the auto sucks the hapless insects into an unexpected death.

If the car is moving at 40 mph or slower, most insects safely flow over it. However, at higher vehicle speeds, they are unable to control their destiny and end up in a fatal crash landing. As a result, windshields are messy as well as hazardous to the motorist who is trying to peer down the road.

The next morning, one can assess the mess, but be wary. The corn borer's misfortune is a boon to another insect. In many cases, yellow jackets have arrived on the scene and are busy extracting the moth carcasses as food for their larvae.

In nature nothing is wasted. Of course, to most folks having yellow jackets swarming around the corn borer carnage on their auto is just adding insult to injury.

Twilight Beckons Insect Jeckyll and Hyde

The best thing about some hot summer days is that they turn into cooler summer evenings. That's when the accumulated heat of day slowly begins to escape the clutches of the earth. It's a time of day when humans appear in backyards to lounge in lengthening shadows, courtesy of the decaying rays of the westbound sun.

Twilight is also the time of day when hawk moths descend on flower beds to sip nectar from awaiting blossoms. Hawk moths are medium to large moths that are strong fliers. It is because of their speed of flight and rapid wingbeat that they are named after that bird of prey, the hawk.

These moths also are called hummingbird moths. Most feed from flowers like hummingbirds and, in some cases, are about the same size as these diminutive birds. They even hover in front of flowers like these birds and use their proboscis to extract nectar from the blossoms.

In addition to being called hawk or hummingbird moths, these insects also are called hornworms or sphinx moths. These latter names are based on the immature form. Larvae have a conspicuous horn protruding from the top of their last segment and are appropriately called hornworms. Hornworms sometimes rear up when disturbed, and this makes them look a little like the Egyptian Sphinx. Thus, some people have called them sphinx moths.

Larvae of hawk moths feed on plants. The most commonly observed of the hawk moths consumes tomato foliage as a larva and is called the tomato hornworm. Most gardeners recognize this insect as a serious pest of tomato plants because of the amount of leaf tissue that it can consume before it crawls from the plant to form a pupa. The insect then emerges from the pupa as a moth and begins to spend its evenings looking for flowers in search of a little sip of nectar.

The tomato hornworm is truly one of the Dr. Jeckyll and Mr. Hyde insects in this world. As a larva, it feeds on and damages our tomato plants, so we consider it a pest and try to get rid of it. As an adult, it is a beautiful and graceful moth that is fun to watch as it gathers nectar from our flower beds.

Most children have at one time or another tried to catch a hawk moth as it hovered in front of flowers in the early evening hours. Most are not successful. After all, in the adult stage a hawk moth is one of the fastest insects known, reaching speeds of nearly 30 mph.

Velvet Wings

1. What was the probable origin of the term butterfly?
2. What is the function of the tent produced by larvae of the tent caterpillar?
3. Why are cutworms called cutworms? What is the common name of the adult cutworm?
4. What is the food of the giant silkworm moths?
5. How is the term wanderer used in reference to insects?
6. In general, how would you distinguish the larva of a moth from the larva of a butterfly based on their hair covering?
7. Why are hornworms so called? What is the common name of the adult hornworms?
8. What moth is said to squeak like a mouse?

CHAPTER 8

Insects in Armor (Coleoptera)

The Beetle

"Ladies and gentlemen, The Beatles!" With these words, Ed Sullivan wasn't introducing insects to his stage, but his famous words are fitting for introducing the largest insect order in terms of numbers of species, the beetles, or in scientific terms, Coleoptera (from the Greek *koleos* meaning sheath and *pteron* meaning wing). This group includes familiar insects such as the lady bird beetles (or ladybugs), the fireflies (or lightning bugs), boll weevils, snout beetles, and scarabs.

Of the 1.75 million species of organisms described on our planet by scientists (Groombridge and Jenkins 2000), approximately 400,000 are beetles (Hammond 1992), which led the late British biologist J. B. S. Haldane to remark that God must have had "an inordinate fondness for beetles" (Gould 1993). That means approximately 1 in 4 species seen on this planet is a beetle, including not only the number of animal species, but plants as well as animals. This proves yet again that insects are the most abundant animals known. Why aren't there more entomologists then?

If beetles are so numerous, then we'd better learn how to recognize the group. The adult beetle possesses two pairs of wings, a hardened forewing, the elytra, which covers and protects the wings used for flight, the hindwings. The forewings fold across the back, meeting at the midline, forming a distinct line down the middle of the back. Beetles have chewing mouthparts in the immature stage and as adults, which can appear somewhat frightening at first sight. The mouthparts of some of the more voracious predators are reminiscent of Death's scythe. They use these mouthparts to feed on other animals, living plant material, or rotting plant and animal matter. Beetles undergo complete metamorphosis and many of their larvae are commonly known as grubs.

A large part of the success of the beetles is their ability to live in almost every habitat imaginable, and many of their common names reflect those habitats. For example, flower beetles live and feed on flowers. Fungus beetles live in the shelf fungus found on trees. Bark beetles live underneath the bark of trees. Similarly, wood boring beetles live in the wood of the tree. Water loving beetles live in ponds, and of course dung beetles enjoy fecal material. There is even one rarely seen beetle that is a parasite on beavers.

Beetles have been around a long time. Because of their very hard exoskeleton, beetle fossils are common. The earliest fossils date back to 250 million years ago (Erwin 1997). Fossils have been found which date to the last ice age, approximately 10,000 years ago, in which it was discovered that the beetle's exoskeleton, made up of the protein chitin, did not mineralize, but remained chitin.

References

Broombridge, B. and M. D. Jenkins. 2000. Gobal biodiversity. Earth's living resources in the 21st Century. World Conservation Press, Cambridge, UK. 246 pp.

Gould, S. J. 1993. A special fondness for beetles. Natural History 102:4–12.

White, R. E. 1983. Peters Field Guide Series. Beetles. Houghton Mifflin Co., Boston, MA.

Flies of Fire

They are one of the true wonders of the insect world: fireflies. To some folks, no warm, summer evening would be complete without the flickering aerial gymnastics of these beetle pyromaniacs.

Many a youngster has spent countless evening hours pursuing the elusive creatures. Much of the fun is in the chase and capture. However, like any big-game trophy hunter, the quarry must come home. Filled with fireflies, a jar provides a barrage of flickering light. A miniature aurora borealis on the windowsill. The intensity of the insect-generated lightning flashes can even keep the successful hunters from falling asleep.

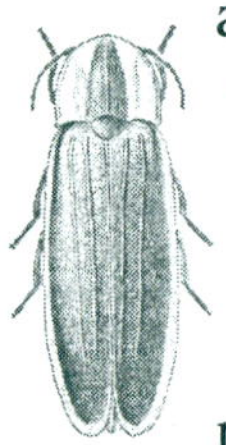

But fireflies don't fly around flashing just to become a prisoner in an empty peanut butter jar. They are flashers with a purpose. Their goal is to electrify one of the opposite sex—to find a mate.

The romantic purpose of firefly activity has not been ignored by the human population. Indeed, the imagination of many a poet has been stirred by fireflies. One of the most interesting poems is that of James Montgomery, who explains the purpose of the firefly's beacon:

When Evening closes Nature's eye,
The glowworm lights her little spark
To captivate her favorite fly,
And Tempt the rover through the dark.

Conducted by a sweeter star
Than all that deck the fields above,
He fondly hastens from afar,
To sooth her solitude with love.

While Montgomery cites the glowworm as his main character, he is actually describing firefly activity. The term glowworm is used most commonly for the larvae of fireflies. These larvae are found in the fall of the year in the vicinity of streams. They do not flash their neon lights but instead produce a constant glow—a real sight on a cool fall evening along some gently flowing brook.

The firefly has been immortalized in the Mills Brothers song "Glowworm." Who among us, young or old, cannot help feeling a spark of romanticism when hearing the lilting strains of this classic:

Glow little glowworm, turn the key on,
You are equipped with a taillight neon.
You've got a cute vest, sparkin' master,
Which you can make both slow and faster,
I don't know who you took the shine to,
I've got a gal that I love so.
Glow little glowworm, glow.

Once again, the insect in the song is not technically a glowworm, but a firefly. But hey, when you're in love, who cares?

Waste Disposal Experts

The ancient Egyptians worshiped them as a god. Barefoot farm boys in denim overalls have admired them for decades. Scientists have studied them. My teenage daughter thought they're gross!

They are popularly known as dung beetles, tumble beetles, or tumble bugs. Technically known as scarab beetles, these insects feed on manure. Not any old manure, mind you, but mammal droppings left in pasture areas. They are specialists in cow-pie disposal procedures.

Mated pairs of tumble beetles dig burrows in the soil, a home for their prospective offspring. Following construction, the tumble beetles begin to provision the home. The food chosen is mammal excrement, frequently cow manure. Once a cow-pie—the fresher the better—is discovered, the beetles must package the food. They make it into a ball—a very convenient shape for transporting—and roll the food home.

It is this habit of rolling the dung ball that gives the insects their common names. Normally, one beetle pushes the ball with its hind feet while its mate pulls while walking backwards. This procedure results in uncertain progress at best. The journey is punctuated with frequent falls and tumbles. But where there's a will, there's a way, and the two dedicated travelers usually make it home.

Once the manure ball is safely in the underground home, the female lays eggs on it. The young larvae hatch and begin to feed on the manure and possibly the fungus that grows there. Dedicated mother that she is, the female beetle frequently remains in the burrow to tend to the brood.

It was the ball-rolling habit of the tumble bugs that no doubt attracted the attention of the ancient Egyptians. In fact, the scarab became the most popular insect in Egypt. The scarab beetle common in Egypt has five terminal segments on each of its six legs. To the ancient Egyptians, this symbolized the 30 days of each month.

The ancient Egyptians adopted the scarab as the symbol for their sun god, Khepera. It is said that they believed that the beetle rolling the dung ball across the earth symbolized Khepera rolling the sun across the sky. Because the sun was "reborn" each day, the beetle became the symbol of rebirth and was associated with the idea of the soul.

The Age of Beetles

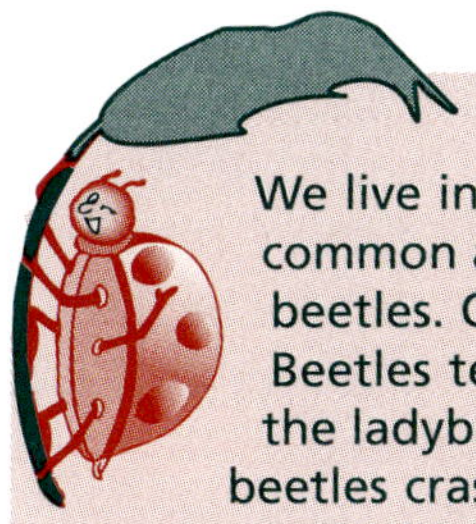

We live in the age of beetles. This idea may puzzle some people, but the truth is beetles are the most common animal present on the surface of the earth. There are well over one quarter million species of beetles. One animal in every four is a beetle!

Beetles tend to keep a low profile, so they're not frequently observed. To be sure, most of us notice the ladybird beetle feeding on aphids on our rosebushes. We may even take note of the sound of May beetles crashing into lighted window panes in the spring and summer. And, on occasion, we fleetingly admire the black beetles that manage to wander into our basements—that is, before we step on them. But unlike their showier cousins, the butterflies and moths, beetles generally go unnoticed.

Beetles are equipped to avoid the limelight. The hard shell that beetles wear as protection allows them the freedom to inhabit the underground of nature. Beetles are often found under things such as leaf litter, tree bark, rocks, boards, and carpets. That hard shell will even allow them to withstand the footfall of a hefty human, at least if the soil is soft.

The name beetle comes from the Old English word meaning "to bite," an early recognition of the rather fearsome biting mouthparts of many beetles. The word beetle is also used to describe a heavy hammer and something that projects, such as a beetle-browed person.

The look of beetles, that indescribable beetleness of this group of insects, was simulated in the shape of that now-famous automobile—the Volkswagen beetle. Some folks have been known to call the Volkswagen version a bug—an entomological error. Real beetles are not bugs, but they are insects.

Of course, any discussion of beetles brings to mind the British rock group of the same name. It is not entirely clear how the four lads from Liverpool arrived at the name the Beatles. However, it has been suggested that they so admired Buddy Holly and the Crickets that they too wanted a name with an entomological connection. They may have changed the spelling to Beatles to reflect the style of music they played, known then in England as "beat" music.

In the early days of the British rock invasion, some people who saw the Beatles and their haircuts—or lack thereof—were probably reminded of beetles of the insect variety. To some their music may have conjured up thoughts of a band of insects. But then life was never meant to be easy, for beetles or Beatles. The success of both groups speaks for itself.

Scarab beetles were frequently put in tombs in an effort to ensure the immortality of the soul of the departed. Nearly 3,000 years later, Roman soldiers adorned necklaces with a scarab form and wore them into battle. This tradition was apparently based on the same idea of immortality that had been accepted by the Egyptians. Necklaces with the scarab design can still be found today.

Dung beetles serve a very practical function, in ancient times and today, that of keeping our pastures free of animal manure. The Australians learned this when large mammals were introduced to their country. You see, Australian dung beetles had not adapted to dealing with manure other than that of the kangaroo, so the Australians had a problem piling up. They solved it by importing "foreign" dung beetles.

Maybe the ancient Egyptians were correct. If dung beetles aren't magical, they are surely essential!

Tigers of the Insect World

To most people, the tiger is a ferocious animal, a blood-thirsty carnivore. So it is in the insect world, which is why tiger beetles are aptly named. They are among the most vicious of insects.

Tiger beetles feed on other insects. They have large sharply-pointed jaws and large eyes, the better to spot their prey.

They capture most of their prey by running. Tiger beetles can run at speeds of up to 60 centimeters in a second, or nearly 1.5 miles per hour. Not so great, you say? Well think about that speed in terms of body length. A tiger beetle is about 1/2 inch long. A race horse about 8 feet long would have to run about 250 miles per hour to cover its body length as speedily as does a tiger beetle!

Tiger beetles like to hunt in open spaces, where speed is useful. Once a tiger beetle catches an insect, its eating habits live up to its name. Tiger beetles seize the prey in their jaws and bang the

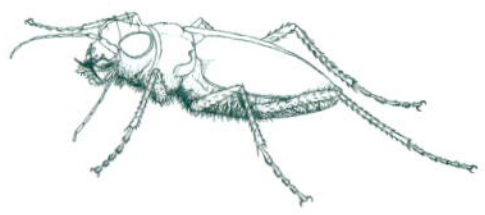

unfortunate victim against the ground until dead. Then the tiger beetle sucks the juices from its prey. For dessert the tiger beetle may just proceed to chew some of the parts of the insect's shell.

The young of tiger beetles are at least as proficient as their parents in procuring food. However, lacking the wings and feet of the adults, the larvae resort to trickery.

Tiger beetle larvae live in the soil where they construct a burrow. A larva climbs to the mouth of the burrow and places its head, which is mostly jaws, out through the opening. When a potential meal passes close by, the larva rushes out, captures the prey and pulls it back into the burrow. Once inside the burrow, the juvenile tiger beetle may retreat to the dark depths of the inner sanctum to dine.

Tiger beetles are not often observed. Primarily because they frequently fly away when people approach. Apparently we are too large to qualify as a potential meal. However, many tiger beetles have beautiful iridescent green and blue colors and are quite showy.

Next time you are walking in an area without vegetation, such as on a dirt path, a sandy beach or even near a log in a sunny spot in the woods, keep an eye open. A few feet ahead you might spy a tiger beetle. With a little patience you might even observe a tiger beetle stalk and capture prey. You will then appreciate why these insects are called tiger beetles.

Insect Flashers

No warm July evening would be complete without the flickering lights of fireflies. Sometimes called lightningbugs, these insects are neither flies nor bugs. They are beetles that have long amazed scientists and amused children with their ability to produce light.

Although scientists have studied the lighting mechanism of fireflies, it is not completely understood. However, the idea has been incorporated in the development of chemical light sticks, such as emergency flares. The light produced by the firefly is a cold light with efficiency in excess of 90 percent. By comparison an incandescent lamp is about 10 percent efficient, which means about 90 percent of its energy is used for heat. Sunlight is about 25 percent light and 75 percent heat.

Fireflies produce light from a substance called luciferin. When luciferin is mixed with oxygen in the presence of an enzyme called luciferase, light is produced. Both luciferin and luciferase are based on the Latin word *Lucifer*, meaning "to bring light." The same name was given to the angel of light, who fell from grace and became Satan.

Scientific names of fireflies also show an association with light. For example, one genus is called *Lampyris*. "Photo" is found in the common genus name *Photouris* and *Photinus*. A species within the genus *Photinus* is named *pyralis*, which is drawn from the Greek root *pyr*, meaning "fire."

Fireflies flash their lights in various patterns—sort of the Morse code of the insect world. The flashing has one purpose: to attract a mate. The males fly overhead, advertising with their mobile neon lights for a mate. Females commonly remain on the vegetation but respond to an attractive male with similar flashing. When both sexes agree that they are the light of each other's lives, the male lands near the female. Then they "turn out the lights" and proceed with the firefly mating game.

Most of the flashing activity of fireflies occurs in the early evening hours, beginning at dusk. By midnight, most of the outbursts of miniature lightning have subsided for another day.

Fireflies are predators, both in the immature and the adult stages. They consume slugs, snails and other insects. This food-procuring activity takes a strange twist in one firefly species. The female mimics the flashing pattern of males of other species of fireflies. When the lovestruck male lands by the female, he becomes not a mate, but a meal. This firefly is known as the "femme fatale" of the insect world.

Could it be that even among the beautiful and graceful fireflies there is a need for truth in advertising?

Fly-by-Night Beetles

Each year, thousands of beetles arrive in the spring to attend Mother Nature's annual "coming out" party for insects. Appropriately, these debutantes are called May or June beetles.

June beetles are well-known to most homeowners and gardeners. In the adult stage, these beetles feed on the foliage of broadleaf plants. In their larval stage, called white grubs, they feed on

Ladybug, Ladybug, Fly Away Home

Some people know them as ladybugs. They're also known as ladybird beetles or lady beetles, which are more appropriate names because these insects are beetles. Lady beetles are bright orange or red in color and frequently have black spots.

There are about 370 species of lady beetles in North America. Most lady beetles are beneficial because they feed on a group of pest insects called aphids. Aphids feed on plants—our garden plants, our crop plants, even our decorator plants. Because of their appetites, the lady beetles can "eat up" an aphid problem. A larva of a lady beetle will consume 200–300 aphids before it becomes an adult. Some lady beetles will lunch on over 500 aphids before laying eggs. That's 15 to 20 aphids per day.

Farmers in the Middle Ages recognized that aphids could destroy their crops. To help with the aphid problem, the ancient farmers sometimes prayed to the Virgin Mary, who is also known as Our Lady. The aphid-eating beetles frequently saved the crop and gained the name of "Our Lady" bug.

Ancient reverence for the insect and the season-ending practice of burning crop stubble led to the familiar nursery rhyme about the ladybug.

Ladybug, ladybug, fly away home,
Your house is on fire, your children will burn!

The ladybug is admonished to fly away home to save her wingless youngsters before they perish in the fire. The rhyme probably also reflects that ladybugs migrate during the fall season. They seek protected places to spend the winter and, at the time, could be found far from "home."

We still recognize the beneficial aspect of lady beetles—a process called biological control—in our gardens and crops. Some gardeners purchase lady beetles by the quart for release. It is important for homeowners considering such a practice to make sure that aphids, which are the food source for the beetles, are present at the time of release. If food is not present, a modern version of the nursery rhyme will occur, and the beetles will "fly away, fly away" to some other home!

the roots of grass. In either case they can cause severe damage to the plants attacked.

So what's a person to do when faced with white grubs in the lawn? Some rely on natural control measures, such as moles and skunks, which feed on the white grubs. Most people, preferring instead to prevent further damage to their lawns, choose to use insecticides. However, chemical control of white grubs is only effective when applied to lawns in early spring or late summer.

These insects have an interesting life cycle. Most species overwinter as mature adults in the soil. In the spring, the adults burrow from the soil and fly about, feed on leaves, mate and lay eggs. Larvae feed on organic matter in the soil. During the winter or under dry conditions, the larvae burrow deeper in the soil. They return to the surface as soil temperature or moisture content increases.

Some June beetles, like the Japanese beetle, spend only one year feeding as a larva. Other species spend two or three years in the larval stage.

Adult June beetles are night fliers. The fact that they fly at all is a minor miracle of the world of nature. These beetles are compactly built, sort of the sumo wrestlers of the insect world. Like all beetles, their front wings are modified into elytra. These shell-like covers protect the insect's body and conceal the hind wings. When a beetle takes flight, it must move the elytra forward and unfold the membraneous hind wings.

June beetles look awkward in flight. However, their takeoffs and flight are almost graceful compared to their landings, which fit nicely into the category of "crash!" Beetles just fold up their wings and drop in. In fact, the poet James Whitcomb Riley once described the flight of June beetles as "bumping along the dusk."

June beetles are attracted to light. Frequently during spring and early summer months, quiet evenings are punctuated with the sounds of June beetles crashing into lighted window panes. Such grounded beetles, apparently no worse for the wear, repack their wings under the elytra and proceed on foot.

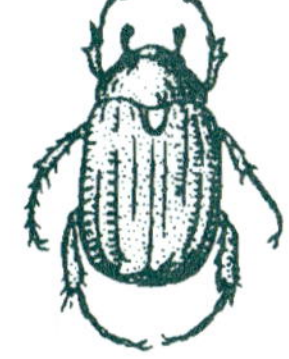

Prowling neighborhood cats have discovered that the ground below lighted windows provides a veritable smorgasbord of crunchy morsels, especially during heavy flights of Junebugs. At least June beetles are good for something.

Just Looking for a Home

People in the cotton industry wish the boll weevil would look for a home somewhere away from the cotton belt. The destructive little pest causes enormous losses in the production of cotton fiber and cottonseed each year.

The boll weevil invaded Texas from its original home in Mexico, and now is residing wherever cotton is grown in the United States except for California, New Mexico and Arizona.

The boll weevil name is widely recognized in America. It was popularized in *The Boll Weevil* song. The first verse of the song says:

Oh the boll weevil is a little black bug,
Come from Mexico, they say.
Come all the way to Texas,
Just looking for a place to stay.

Technically, this insect is not a bug. Weevils are beetles, sometimes called snout beetles, because they have mouthparts on the end of a snout. Boll weevils use the snout to feed on the cotton bolls which causes this fruiting part of the plant to fall and reduce the yield.

The second verse of the song is:

The first time I seen the boll weevil,
He was sitting on the square.
The next time I seen the boll weevil,
He had all his family there. . . .

How true it is. The square is the flower of the cotton plant and is a favorite feeding site for the weevil. The boll weevil female lays eggs in the feeding punctures, up to 250 per female. The young grubs then feed in the developing boll.

The rest of the song deals with the damage caused by the boll weevil and the economic straits of the affected farmers. According to the song:

The merchant got half the cotton,
The boll weevil got the rest.
Didn't leave the farmer's wife
But one old cotton dress,
And it's full of holes, and it's full of holes.

In spite of the damage caused by the boll weevil, the insect is a hero in Enterprise, Alabama. So much so that in the center of town stands *The Boll Weevil Monument*. The monument was erected in recognition of the role that the boll weevil played in causing crop diversification in the area. Peanut production was introduced, and the economic climate of the area became more stable.

Even major pest insects, like the boll weevil, sometimes provide opportunities for folks other than entomologists.

Bombardier Beetles

Bombardier beetles are fittingly named. Historically, bombardiers were people who waged war by lobbing a variety of objects at their foes. Bombardier beetles don't drop bombs, but they have mastered the art of chemical warfare.

When these insects are attacked or disturbed they live up to their name. The bombardier beetle ejects a secretion composed primarily of chemicals known as quinones. The quinones are not stored in the insect. They are produced as needed in a complex reaction that would make any chemistry teacher proud.

The reaction, which takes less than a second, produces heat and gaseous oxygen. The gas propels the mixture from the insect with a popping sound. The temperature of the explosion has been shown to reach 100 degrees Celsius, the boiling point of water. Consequently, any creature unlucky enough to raise the ire of a bombardier beetle is sprayed with a hot chemical mixture.

Bombardier beetles are not just mad bombers. They are excellent marksmen—or should we say "marksbeetles"?—who direct the spray precisely. The gland that produces the defensive chemicals is located at the tip of the abdomen. The abdomen works like a movable turret that is pointed in the direction of the attack. If an ant attacks a beetle's leg, for example, the beetle "bombards" in that direction.

The chemicals in the beetles' spray are very effective. A toad attempting to make a meal out of a bombardier beetle is rewarded with a mouth full of hot chemicals. The amphibian spends the next few moments doing a toad's version of "yuk," probably seriously rethinking a life devoted to eating insects.

Bombardier beetles really are pacifists in their little insect hearts and will only live up to their name as a last resort. When disturbed, they try to avoid trouble by running away. If such an evasive technique does not work, they pull out all the stops and employ their chemical weapons. For these beetles "bombs away!" is not just a cute saying, it is a matter of survival.

Sexton Beetles

Years ago every church had a sexton. As under-officers of the church, sextons had several jobs. They rang the church bell; took care of church property, including the graveyard; and dug graves.

There are a group of beetles, called sexton beetles, who, like their church namesakes, are gravediggers par excellence. Also known as scavenger beetles or burying beetles, sexton beetles make their ecological living by burying the carcasses of small animals as a food resource for their young. They are carrion eaters.

The rather grisly life of the sexton beetles also is reflected in the scientific name of one species. *Necrophorus marginatus* has a name that can be roughly translated as "the body carrier." The sexton beetles are attracted to a carcass by the odor of death. If the dead animal happens to be on soft earth, the beetles bury it where they find it.

A dead mouse on a hard surface, however, presents a challenge to these six-legged gravediggers. The solution is simple: just move the carcass to a place where digging is easy. Doing it is another matter, at least if the insect is moving a dead weight several times heavier than it is.

Sexton beetles have solved the problem, they crawl under a carcass and do the insect equivalent of standing on tiptoes. This process scoots the car-

Here a Grubworm, There a Grubworm

Editors Note: In the fourth paragraph, the reference to "Junebugs" is correct as is, even though it may not match your dictionary. Scientifically speaking, since a Junebug is NOT a true bug, the words "June" and "bug" are combined into one word. Most dictionaries list this reference as two words. A future column will address this subject in more detail.

Most lawn owners and home gardeners are eternal optimists, at least in the springtime. During the first warm days of spring, these creatures emerge from their domiciles armed with rakes, spray cans and fertilizer bags. With the accumulated energy of winter relentlessly driving these people onward, they prepare to battle Mother Nature over the appearance of a bit of real estate called the lawn. The eternal springtime dream of an insect-free lawn is quickly shattered through the specter of a grubworm infestation.

Grubworms, sometimes just called white grubs, are the scourge of the green-lawn enthusiast. These insects are the larvae of beetles known as Junebugs or May beetles. Japanese beetle larvae are also called white grubs.

These insects deserve to be called grubs. They spend their life underground grubbing around for food. Some, including an insect called the masked chafer and the Japanese beetle, spend only one year in the larval, or grub, stage. Other grubs may spend two or three years underground.

Grubs feed on plant roots, especially the roots of grasses, such as those found in lawns. Most feeding occurs during the warm months of the year. During cold or dry periods, the grubs burrow deeper in the soil and stay there until the soil returns to desirable temperatures or moisture.

Grub feeding does horrible things to the grass plant. They deprive the plant of its ability to procure water and nutrients, so it dies. This, of course, results in brown spots in the lawn.

Neighbors have the uncanny ability to notice grub-induced brown spots in lawns and immediately begin to talk: "Guess who has grubs?" Of course, the social disgrace of having grubs is more than most lawn owners can take, so they do what they can to solve the problem.

Where there are grubs, there are also grub eaters. Moles, blackbirds and skunks are grub eaters of the first order.

While moles and skunks can rid a lawn of grubs, the solution might be worse than the problem. Indeed, no one is certain how many skunks per lawn it takes to eliminate white grubs. In addition, this control method carries with it the risk of exceeding the clean air standards of some cities.

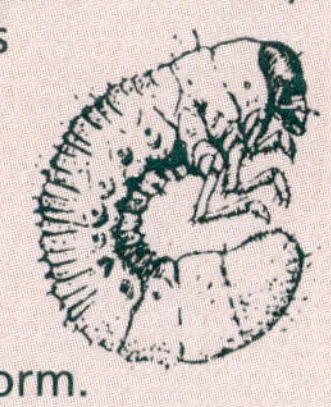

Insecticides are useful tools in dealing with the grubs. But applications need to be properly timed, which is tricky. Also, the use of insecticides carries with it some environmental hazard.

There are also other methods. Some folks have tried walking around on the lawn with spiked-shoe, lawn "aerators" attempting to spike the grubs. No one is sure if this method is effective, but it gives the owner the satisfaction of having gotten even with at least one grubworm.

cass forward millimeter by millimeter. After every to or three shoves, a beetle will merge from under his prize carcass and scurry around to survey the progress being made. It then scampers back to continue the working frenzy.

To sexton beetles, speed is of the essence in burying a carcass once it has been discovered. They need to cover it up before other carrion-eating insects, such as flies, get a chance to share in the feast. And what a feast it will be! Once the carcass is buried, the sexton beetles will lay eggs on it. The eggs hatch quickly, and the young larvae that emerge will dine to their hearts' content on the repast provided by their parents.

Historically, humanity has looked down on the job of gravedigger. Most of us also tend to put the life of the sexton beetle under the heading of "yuk." Theirs may be dirty little jobs, but ecologically speaking someone or some insect has to do it!

Eating Beetles Is No Picnic

Let's have a picnic! Eating in the great out-of-doors ranks right up there with swimming and going to the ball game as things to do in the summer. We even have special equipment named for the occasion—picnic baskets and picnic tables.

You can't have a picnic unless someone shows up with the food. The food attracts not only ants but another insect named especially for the event, the picnic beetle. This uninvited picnic crasher is a black beetle with four yellow dots. It's about the same size as a house fly and is sometimes called a fungus beetle or sap beetle.

These common names all reflect the food habits of the picnic beetle. The insect in both the adult and immature stages feeds on decaying plant material or the fungus that grows in such situations. It commonly can be found around plants that have been damaged. Some entomologists have suggested that the insect may be more common in years when corn is badly damaged by corn borers. The damage to the plants provides breeding sites for the beetles.

In fact, picnics are frequently smorgasbords of damaged and fermenting plant material. Potato salad, cole slaw, cucumbers and relish trays must seem like heaven to hungry picnic beetles. So when these beetles discover our picnic they drop in for a bite.

These beetles drop in literally. As they fly about searching for food, the odors from the picnic dishes are a signal that dinner is served. The beetles then fold their wings and fall from the sky. Sometimes they are on target and fall right into the dish of choice—theirs, not ours. Sometimes they crawl around to find it.

Always, their presence is disconcerting to picnic-goers who want to eat outdoors but don't want to share their food with outdoor creatures; or worse yet, who shudder at the thought of inadvertently devouring one of the little creatures in a spoonful of potato salad.

It's not easy to keep picnic beetles from crashing a picnic. Some people have been known to place jar lids of beer laced with a dash of insecticide in the vicinity of the picnic-to-be a day prior to the event. The picnic beetles are attracted to the fermented plant sap known as beer and get a dose of insecticide.

Most of us have just learned to live with picnic beetles. We, of course, eat more carefully when they are around. After all, sharing your picnic with an insect is one thing, but eating an insect in your potato salad is, shall we say, "no picnic!"

Last Glow of Summer for the Insect World

They are the dying embers of the blazing fire of summer insect life. Glowworms, the immatures of fireflies, provide the finale, the curtain call for the season of insects.

Glowworms, like adult fireflies, have the ability to produce light. Unlike their parents, however, glowworms do not flash. They produce a constant glow, which is the basis for their name.

Many people have never basked in the glow of a glowworm. Unlike adult fireflies, glowworms do not produce aerial displays that are hard to miss. Glowworms, like all immature insects, are wingless. They crawl on the soil surface or live in the soil. Generally, they are out of sight and, therefore, out of mind for most humans.

Most glowworms are associated with moist environments. Swamps, creek banks and edges of ponds are prime glowworm habitats. However, large concentrations of glowworms have been spotted in drier areas, such as in the leaf litter of an oak woods.

Glowworms are predators. They feed mainly on slugs and snails but also will make a meal out of other soft-bodied creatures like caterpillars.

Without their lights, glowworms are hard to spot. They are brown to black colored and blend in nicely with moist soil. To most people, glowworms resemble those creatures we know as pillbugs or sowbugs. Pillbugs are not insects but are common creatures found in decaying plant material, such as compost piles or mulched flower beds.

If you want to see glowworms, you need to head out to a creek or pond. Dark nights work best since the lights are difficult to see in bright moonlight or, for that matter, under a street light.

Patience is a virtue when it comes to glowworm viewing. Glowworms are somewhat shy creatures, so when they are disturbed they douse their lights. A person walking through their habitat will prompt a mass extinguishing of the little lanterns.

So, if you go looking for glowworms, walk into the area and stop. As you stand quietly, focus your eyes on one spot. Stare at the spot as you might stare at a 3-D picture. Suddenly, the glowworms will reward you with first one and then other little blue-green lights. Once you have spotted the light, you will find them everywhere.

One of the great scenes for nature lovers is to witness literally thousands of glowworm lights lining their favorite stream bank on a dark night. It is an awesome sight.

Generally, scientists cannot explain the function of the light for glowworms. One suggestion is that the light allows them to mark their territory. It also has been suggested that the light works as a reverse-flash protective device. The insect is easy to see with its light on but when disturbed, it turns the light off and moves. This makes it difficult for a predator to find it again.

Whatever the biological reason, the little lights of immature fireflies are another miracle of nature. Glow little glowworm, glow!

Fireflies Bring Out the Romantic Poet in Most of Us

The American poet Ogden Nash, with his characteristic wit, once proclaimed "The firefly's flame is something for which science has no name."

True.

But science has named the chemicals that produce the light. These chemicals are called luciferin and luciferase. Luciferin is the substrate, the fuel for the fire. Luciferase is an enzyme that makes the reaction go—sort of like pouring kerosene on a fire.

So the firefly injects oxygen into the system and produces light. It's the light production that suggested the name for luciferin and luciferase. These chemicals are named after Satan, the fallen angel of light.

Chemistry aside, almost anyone who views light production by the soft beetles we know as fireflies or lightning bugs is amazed. So taken are we by the flashes on warm summer evenings that many people wax poetic about the sight.

The poet Lowell described fireflies as "flickering gold" and "fitful heat-lightnings." Henry Wadsworth Longfellow called fireflies "white-fire creatures" in "Hiawatha's Childhood."

Jones Very in his poem "Fireflies" described the insect "Like sparkles glancing to and fro among the new-mown grass." Cowper also chose spark to describe firefly light in his poem "The Nightingale and the Glowworm."

A. L. Ruter Dufour called a firefly "a bright star-child." Robert Frost thought the analogy to stars might apply to fireflies. But he declared in "Fireflies in the Garden" that the insect was only an "emulating fly" of real stars. He admitted that the firefly does "Achieve at times a very star-like start."

The popular Mills Brothers song of the '50s, "Glowworm," featured several analogies for the firefly. The insect is called a "fly of fire," which "glows like an incandescent wire." To further the electricity analogy, the song suggests the insect should turn on the AC and the DC!

Not to omit the other popular common name of this insect, the song mentions "bug of lightnin" which was equipped with a "taillight neon." The insect is also referred to as "sparkin' master" in the song.

Both poets and songwriters have immortalized the firefly in their works. But not lost on these wordsmiths has been the real reason that the firefly produces light. To attract a mate! Yes, all of that glittering, sparking and pretending like they were stars or lightning is designed to attract the attention of the opposite sex.

So in the song "Glowworm" it is no surprise that the lyrics go from fireflies to humans. For instance the lines:

"See how the shadows deep and darken,
You and your chick should get to sparken',
I've got a gal that I love so.
Glow little glowworm, glow."

Other songsters as diverse as Tony Bennett, The Burr Oak Ensemble and Fleetwood Mac also have recorded songs with a firefly theme. After all, most of us who have ever watched firefly displays over a field in summer time have got to appreciate the romantic side of the display.

Insects in Armor

1. What American music group is said to have inspired the Beatles to choose an insect name for their group?
2. According to the Old English language, the word which is the basis for beetle had what meaning? Why is this appropriate for this group of insects?
3. What is the name of the substrate and enzyme used by fireflies in light production?
4. What is the ecological role of scarab beetles?
5. What is the femme fatale of the insect world?
6. How would the food habits of tiger beetles be described?
7. Where did the ladybug get its name? What is the food of this creature?
8. What is the food of immature Junebugs? What are they called?
9. What was the original home of the boll weevil?

CHAPTER 9

Flies, Flies, Flies (Diptera)

"Shoo fly, don't bother me!" How many times have you said that? Flies, including mosquitoes, gnats, and midges, can certainly be a bother, but the majority of them want nothing to do with humans. They belong to the Order Diptera (from the Greek *di* meaning two and *pteron* meaning wing). There are over 100,000 species described, with nearly 20,000 occurring in North America. Their size can range from 0.5 mm (the size of a period on this page), to 50 mm (the height of a film canister). Flies occupy many jobs, or niches, in their environment, including scavengers, predators, parasites (both helpful and harmful, from a human perspective), and pollinators. They also wear the hats of pests, both of plant and animal, and vectors of disease, such as malaria, yellow fever, and sleeping sickness. With all of the "flies" in the world, how is it possible to tell a real fly? It's all in the name! Stoneflies, mayflies, and dragonflies are just a few of the many insects that have the word "fly" within their name. These are not true flies, having "fly" attached to their name. On the other hand, true flies have "fly" detached from their name (i.e., hover fly, house fly, and deer fly).

Besides their name, flies have morphological characters that distinguish them from all other insects. Adult flies have only one pair of membranous wings, with the second pair being modified into balancing organs, called halteres. These organs appear to be little balls at the end of stalks coming off the thorax. The mouthparts of most adult flies are used for sponging moist food, yet some are specialized for piercing-sucking or cutting-sponging. The least favorite of all flies, the ones that bite and transmit disease, have the latter mouthpart types.

Diptera go through complete metamorphosis. Eggs hatch into larvae, commonly referred to as maggots. Maggots live in a wide variety of moist environments, they are without legs, and they often lack a defined head, making it difficult to tell one end from the other. The pupal stage often develops inside a puparium, consisting of the last exoskeleton shed by the larva. Flies are probably the least respected of all the insects, but without their recycling efforts, our world would be a pretty stinky place, to say the least!

Doggers of Civilization

Of the world's pest insects, none are more prominent than flies. These "two-winged" insects have been called "doggers of civilization." And for good reason.

Of the 10 plagues of ancient Egypt, two were caused by flies. Fly-borne diseases are said to have hastened the decline of ancient Athens and Rome. Sleeping sickness, a disease transmitted by the tsetse fly, long delayed the civilizing of Africa.

Ancient Semitic people recognized a deity known as Ba'al Zebub, the god of the flies. It might have been wishful thinking, but these people thought that Ba'al Zebub would defend them against flies. The Philistines of the era apparently couldn't depend on Ba'al Zebub. They appointed a fly control officer. No one knows how this first government-sponsored program for insect control worked out.

Songwriters have recognized the importance of flies as pests. "The Blue Tail Fly" is about a species of horse fly that has a vicious bite. The song became so popular that it was number one on radio's hit parade of songs in the late 1940s.

Many folk sayings exist that relate to flies and fly behavior. Most people prefer not to have "a fly in the ointment," but other lessons can be learned through the less common sayings of our ancestors. For example, a lesson on the value of even the smallest thing, "Every fly has its shadow," on hard work, "Flies are busiest about lean horses," on common sense, "Even flies won't light on a boiling pot," on overdoing it, "Cover yourself with honey, and the flies will have at you," on watching what you say, "No fly gets into a shut mouth," and on watching your step, "If the fly flies, the frog goes not supperless to bed."

Flies also have a reputation as weather predictors. Flies collect on the screen door just before a storm. Flies, it is said, "bite sore when there is a good chance for rain."

Modern civilization is still fighting flies. Flies are annoying us and stealing our blood and that of our animals and we can't entirely stop them. To be sure, we have window screens, flypaper, and flyswatters. We even have insecticides, but we still have flies.

The most common of all flies is the house fly. Even its scientific name—*Musca domestica*—indicates how closely it lives with mankind. While the house fly can't bite with its sponging mouthparts, it has been implicated in disease transmission. These flies can carry germs that cause dysentery, diarrhea, and food poisoning.

The house fly, in spite of all its imperfections, appears polite. The little beast always cleans every one of its six feet before tromping around in the mashed potatoes and gravy on your plate.

Justice and the Blue-Tail Fly

The Southern folk song "Jim Crack Corn" relates a mystery of sorts. The song recounts a death plot suitable for the popular TV show *Murder She Wrote*.

In case you've forgotten, the song chronicles the demise of "master." He was rather unceremoniously unseated from the saddle of his pony and tumbled in the ditch. He died. A jury deliberated the crime, and the verdict was that, not the butler, but the blue-tail fly was the guilty party.

Blue-tail fly is a common name for one group of insects also known as horse flies. Adult horse flies and their insect cousins, the deer flies, are blood feeders. Neither insect feeds exclusively on the animal for which it is named. Almost any mammal is fair game for these bloodthirsty flies.

These insects are known for their painful bites and their persistence in obtaining a meal. Anyone who has been bitten or repeatedly dive-bombed by a horse fly or deer fly will attest to those attributes. Fear of the insect's bite and buzzing causes threatened animals to behave erratically. Fly nets were probably developed for use on draft horses to protect them from this group of flies.

Horse flies and deer flies are frequently found in the vicinity of wetlands. Larvae of these insects live in saturated soil. However, adults are strong fliers and are frequently found far away from breeding sites.

Farm kids have always been fascinated by horse flies and have been known to capture the pesky insects. Of course, a horse fly doesn't make a great pet, but they can be used for other purposes. For example, children have been known to "research" the behavior of a headless horse fly. Such decapitated insects will, with proper provocation, fly. Flights of these unguided insect missiles usually end with spectacular collisions with immovable objects, such as trees, fence posts and, best of all, barns.

Such apparently barbaric behavior on the part of kids toward horse flies likely arose as an act of revenge—for having been bitten while skinny-dipping in the creek or pond.

Once again, the jury has convicted a habitual offender—the blue-tail fly.

Seeing Cluster Flies? It Must Be Spring!

Ah, the signs of spring. Robins hopping on the lawn, crocuses emerging from the ground, and spring peepers singing in water-filled ditches along the roadside. An even better harbinger of spring, though, is flies bouncing off window panes.

In the poem "Spring," Oliver Wendell Holmes penned these lines:

The housefly, stealing from his narrow grave,
Drugged with the opiate that November gave,
Beats with faint wing against the sunny pane,
Or crawls, tenacious, o'er its lucid plain.

The fly that Holmes describes is not the house fly, but a similar-looking fly known as the cluster fly. Cluster flies are parasites. In their immature stages, cluster flies feed on earthworms. In the adult stage, they hibernate to while away the long, winter hours in temperate regions of North America.

Cluster flies prefer a nice cozy place in which to hibernate. So each fall, millions of the little buzzers invade buildings, barns, garages and houses. Now most people don't mind sharing their barns or garages with a few flies, but it's a different story when it comes to houses.

The cluster flies normally find their way under eaves and into chimney chases and attics, where they settle down for a little serious hibernation. However, the increasing minutes of sunlight associated with late winter days frequently results in warm attics.

The increased temperatures do not go unnoticed by our slumbering parasitic fly. It assumes that spring has sprung and begins to search for some earthworm on which to lay eggs. However, the return journey to the great out-of-doors frequently results in a slight detour into the interior of the house.

Once indoors, the creatures try to fly out through windows. While window panes do a good job of keeping insects out of our houses, the glass also does a fine job of keeping flies in. The hapless creatures just crawl around or, as Holmes states, "Beat with faint wings" against the pane.

Of course, the presence of flies makes most homeowners a bit fidgety. Some run for the good ol' insecticide spray can. Others just fire up the vacuum cleaner. A few engage in hand-to-tarsal combat with the unwelcome invaders. Regardless of the elimination approach, there always seem to be more. In fact, hibernating flies can sometimes be several inches deep in corners of attics.

Let's hope that most of the cluster flies can find their way out as easily as they found their way in!

Gnats, Gnats and More Gnats

When you come within a gnat's eyelash, you're close. In fact you can't come much closer, short of winning.

April Showers, May Flowers and Mosquito Blood Banks

April showers bring May flowers—or so the old saying goes. But there is a downside to those spring rainfalls. They are also partially responsible for the summer's mosquito crop.

Those night-flying, bloodthirsty mosquitoes live as immatures in standing water. So rains that provide water to nourish the spring blossoms also fill potholes and low-lying areas in woods and back yards that become breeding sites for mosquitoes.

There are lots of different kinds of mosquitoes. More than 100 species live in the United States. They occur from Alaska to the Gulf Coast states; they thrive from the marshes of the eastern part of this country to the deserts of Arizona. And all of them have at least two things in common: Female mosquitoes need a blood meal to produce eggs, and their immatures live in water.

As a general rule, the female mosquito lays her eggs during May and June. She lays them one at a time or in groups of 100 to 400 on the surface of the water. In one to three days, the eggs hatch into what is called a wriggler. The wriggler gets its name from the motion it makes as it moves through the water.

Wriggler mosquitoes do not get oxygen from the water as do other insects that live in the water. Rather, they are air breathers. They get air through a miniature snorkel, a siphon, that is attached to the rear end of the insect. The wriggler breathes by coming to the surface and sticking its snorkel through the surface of the water. The fact that the wriggler has to get air from the surface lead to a method of control in which the water surface is covered with a film of oil. The oil prevented the young mosquito from getting air, and it suffocated.

Mosquito wrigglers feed on microscopic plants, and in a week to 10 days they molt into the pupal stage. In this stage, they are shaped like commas and stay near the surface of the water with their breathing tube in contact with the air. When disturbed, they will leave the surface in a sort of tumbling motion and are sometimes called tumblers.

In two to three days the tumbler becomes an adult mosquito. The skin of the tumbler splits down the back, and the winged adult climbs onto the skin and uses it as a float until it is ready to fly.

The newly emerged adults use their wings to continue the population of mosquitoes. The males fly off to find females, and the females, once mated, fly to find a host to supply a blood meal. That is the part of the mosquito's life cycle that most of us dislike. That is when we are asked to donate a little blood so the female mosquito can produce another generation of eggs. Most of us would just as soon not contribute, but the mosquito doesn't ask. She just helps herself!

How close is a gnat's eyelash? Well it's so close that it's not measurable, because gnats don't have eyelashes! But if these little insects did have eyelashes, they would be too small to see and that is the basis for comparison.

Gnats are small flies, and some of those babies bite. They even take their name from their biting habit. The word "gnat" is based on the Greek word *gnathos* or "jaw."

There are many kinds of gnats. Buffalo gnats, also called black flies, are definitely worth swatting since the females are blood-suckers and vicious biters. Anyone in the northern areas of the United States during the summer can understand the problems these insects cause. Both humans and livestock are tormented by female black flies as they attempt to obtain a blood meal. Immatures of these insects occur in streams. The adults swarm in groups, hence the name, buffalo gnats. Bicyclists and joggers can attest that running or riding through a swarm of buffalo gnats is something that should not be done with a wide-open mouth!

Fungus gnats are slender mosquito-like insects. They are found in damp places where decaying vegetation abounds. These insects frequently occur in mushroom cellars and sometimes emerge from the soil of potted plants. The presence of fungus gnats in a home normally receives less than rave reviews from homeowners who suddenly discover these insects flying around the living room.

There are also root gnats, gall gnats and wood gnats. And there is a small fly sometimes referred to as an eye gnat. This insect seems to be attracted to animal secretions and is particularly attracted to eyes. Because they are attracted to eyes, they are sometimes vectors of a disease of humans and animals called pinkeye.

To most people, a gnat is a gnat, and something to be avoided. Indeed gnats swarming around your face is a bit disconcerting at best. The next time you see a gnat, try and spy its eyelashes. Then and only then will you truly appreciate the closeness of the event that was won "by a gnat's eyelash."

Flying Syringes

Mosquitoes have been called living, flying syringes. A fitting description.

Technically a kind of fly, the mosquito has long been famous for its bite. The ancient Greek philoso-

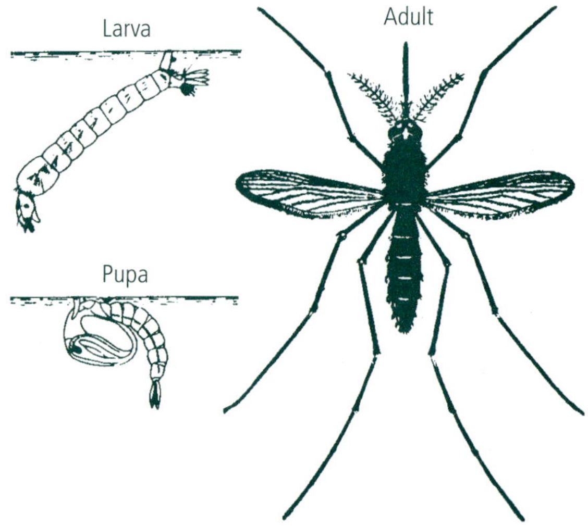

FIGURE 9.1. *Mosquito Life Cycle.*

pher Aristotle, a student of insects, concluded that, "Four-winged insects have the sting in the tail, and the two-winged ones have the sting in the head." Aristotle was, of course, talking about bees and biting flies, probably mosquitoes.

Mosquitoes need to obtain blood for egg production. Therefore, only female mosquitoes bite. The males are nectar feeders. Because of their need for a blood meal, female mosquitoes are notorious carriers of disease. Diseases transmitted by mosquitoes include malaria, yellow fever, dengue fever, elephantiasis and encephalitis.

The relationship between mosquitoes and disease transmission has been an important area of study. Walter Reed, a U.S. Army surgeon, gained fame by leading the effort to control the yellow fever mosquito in Panama. This effort allowed the construction of the Panama Canal. Yellow fever was one of the reasons the French were unable to complete construction of the canal.

The American poet Henry Wadsworth Longfellow suggested a relationship between mosquitoes and malaria in "Hiawatha." And the relationship between swamps and the disease was noted also in the line:

He, the mightiest of magicians,
Sends the fever from the marshes.

This was written some 40 years before it had been established scientifically that the mosquito was a vector of malaria.

Mosquito larvae live in stagnant water, thus the relationship between swamps and mosquito-borne diseases. Many common names of mosquitoes, such as tree-hole mosquito and rain-barrel mosquito, suggest the location of their larvae. One of the most common methods of mosquito control is removing possible larval breeding sites, including the draining of swamps.

Of course, the very best and most modern of mosquito control activities will not be completely successful. So the next time you are being pursued by a hungry mosquito, watch what you say. That mosquito is bound to be a lady!

Flies, Flies, Flies

1. What is the adult food of horse flies and deer flies?
2. Why do mosquitoes need a blood meal?
3. What disease, transmitted by mosquitoes, delayed the construction of the Panama Canal?
4. What is the larval food of cluster flies? How do cluster flies overwinter?
5. What word based on the Greek word meaning jaw is used to describe a group of small flies?

CHAPTER 10

The Socializers (Hymenoptera)

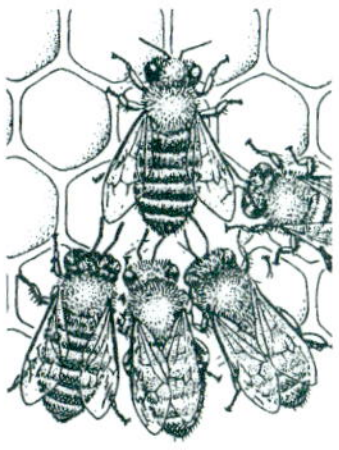

Hymenoptera

The Hymenoptera (bees, wasps, and ants) are an important insect order due to their shear numbers and mass. Bert Hölldobler and E. O. Wilson, the famous myrmecologists (that is a fancy way of saying they study ants), described that C. B. White, a British entomologist, calculated that the number of insects alive at any one time on the earth was one million trillion, and one percent of that are ants, which means that there are ten thousand trillion ants at any given time. Fortunately, they are not all marching toward our picnic baskets! Hölldobler and Wilson (1995) calculated that if the average worker ant weighed from 1–5 milligrams, the weight of the ten thousand trillion ants would be about as much as the weight of all the humans combined.

The Hymenoptera includes 108,000 species of sawflies and horntails, bees, wasps, and ants. The sawflies and horntails differ from the other Hymenoptera in that the connection between the thorax and abdomen is broad, unlike the distinctly narrow junction found in the bees, wasps, and ants. Many of the familiar social insects are members of this order including the honey bees, bumble bees, ants, and paper wasps. But not all the bees and wasps form colonies. In fact, the majority live a solitary existence. Bees and wasps play a critical ecological role in that they are important pollinators of fruit trees, field crops, and almost any other flowering plant.

The Hymenoptera differ from the true flies (Diptera) in that they possess two pairs of membraneous (or transparent) wings, with the forewings significantly larger than the hindwings. Unlike those of the more "primitive" groups such as the dragonflies and damselflies (Odonata), the wings of the Hymenoptera lack extensive venation. In some cases, especially for the very small parasitic wasps such as the fairy flies or trichogrammitid wasps, which can be less than 1 mm in size, there may be only one vein present. The bees and wasps differ in that the bees are generally hairy creatures and the wasps are bald. The long, multisegmented antennae provide another means of identification for this group. The antennae of the ants are unique because they are "elbowed," or angled in a fashion similar to that of the human arm when bent at the elbow. The mouthparts generally are of the chewing type, but some members, particularly the bees, also possess sucking mouthparts which allow them to feed on nectar. They undergo a complete life cycle, and the immatures may be well cared for, as is found in the ants or honey bees, or they may be given a single meal when the egg is layed and left to fend for themselves, which is the case for the solitary bees.

References

Hölldobler, B. and E. O. Wilson. 1994. Journey to the ants. A story of scientific exploration. Belknap Press, Cambridge, MA.

Spring Hive-Cleaning

As days begin to grow longer during late winter and early spring, housekeepers everywhere begin to think of that annual ritual of tidiness called spring cleaning. So it is with honey bees.

Honey bees should certainly win the Good Housekeeping Seal of Approval for their efforts in keeping their hives spotless. However, their early spring efforts might draw the ire of environmentalists who are concerned about airborne pollution.

Honey bees clean their home and empty their digestive tracts by flying from the hive and releasing the refuse while on the wing. What goes up must come down, but the bees are less concerned about the landing site of their airborne sewage than with the tidiness of their home. Consequently, bees' cleaning efforts sometimes result in spots on houses or automobiles. This, of course, does not go unnoticed by the human owners of such items.

With bees, as with people, the call of nature is sometimes a pressing proposition. Bees, like all insects, are cold-blooded. Thus, cleansing flights are confined to times when air temperatures will allow. Sustained flights of honey bees occur when the temperature exceeds 50 degrees Fahrenheit. So during cold periods, waste material begins to build up in the bees and in their hives.

When the first warm period comes around, the bees leave the hive like planes launched on a bombing run from an aircraft carrier, but in this case, the mission is not secret. As temperatures rise, they also remove other trash from the hive. Dead bees, for instance, are rather unceremoniously pitched from the mouth of the hive—the insect equivalent of burial at sea.

Many people try to avoid having their possessions bombed by bees on a cleansing flight. When dealing with bees and their unauthorized deposits, an ounce of prevention is required. It is easiest to move the prized possession out of the line of fire, especially on the first warm day following a cold spell.

However, if bees get to the car before you do, prompting a trip to the local car wash, just be thankful that honey bees are smaller than elephants.

Hornet Artisans

Once again, Mother Nature has completed her age-old ritual: the shedding of leaves. Each year as winter approaches, deciduous trees in temperate climates everywhere drop their excess baggage. Those leaves, like lovers scorned, float aimlessly earthward.

But the annual departure of the leaves reveals more than just bare, nurturing limbs. Hornet's nests, hidden from view by summer foliage, suddenly stand naked against the winter sky. Hornet artisans build and maintain their homes for months. Then in the fall, like the leaves that had concealed it, the hornet's nest is left to the wiles of nature. It will soon be gone.

Bald-faced hornets—their name comes from the white spot between their eyes—are engineers of the highest order. They construct their globe-shaped nests from a paper-like material. The hornets collect wood from any available source, and by chewing and mixing it with saliva, they produce the nest material. Because these insects frequently use different sources of wood, the nest may appear multicolored.

As the summer progresses, the nest is enlarged to accommodate a growing family. By summer's end, many nests are basketball-sized or larger.

Bald-faced hornets pack a powerful sting, and few things are madder than a mad hornet! However, most hornets are quite docile, at least when away from their nests. However, they are willing, even anxious, to sting in defense of the home.

Hornets feed protein to their young. They are active hunters and can frequently be found around homes in pursuit of flies. They have even been known to land on people after mistaking a button or spot on clothing for a food item.

Hornets are beneficial because they prey on pest insects, but some people might say the greatest benefit of bald-faced hornets is the decorative value of their nests. Indeed, many hornet nests have ended up hanging over the mantel of a fireplace. Procuring such a decoration from the wild carries with it the inherent risk of introducing a few hornets into the house.

Each fall bald-faced hornet queens mate and seek shelter in woods to overwinter. The next spring, the queen starts the process over. The remaining workers stay with the nest until they are killed by freezing temperatures.

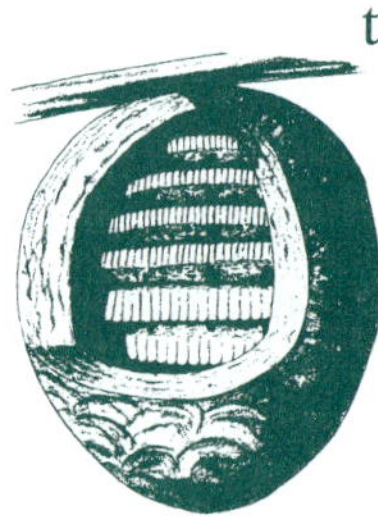

Of course, the rather fragile nest must be maintained constantly, and a nest will not last long when exposed to seasonal winds and marauding animals. Therefore, a good nest for decoration must be procured after the tenants are deceased but before it is blown to pieces. That, my friends, is where the rub—or rather the sting—comes in.

The Wisdom of Ants

Ants can be a nuisance. Especially at picnics, in kitchens or even "in someone's pants." In spite of the problems ants cause, we humans have come to admire the little creatures. It's their work ethic that we so revere.

The industry of ants, their widespread distribution and large population have provided ample in-

On Bees and Stingers

One of the more well-known characteristics of insects of the order Hymenoptera is that they sting. Not all Hymenoptera sting, but those that do certainly get our attention!

The stinging Hymenoptera include insects of three groups: bees, wasps and ants. Most people recognize that these insects include some species that live in groups called colonies. Insect colony dwellers are social insects. A hive of bees, a wasp's nest or an ant hill are frequently encountered examples of social insect colonies.

All ants live in colonies, but many bees and wasps do not. Hymenoptera that shun the social lifestyle are known as solitary species. Most solitary bees and wasps can sting but generally do so to paralyze prey, such as caterpillars or spiders, as food for their offspring.

The social insects can be downright unsocial when it comes to their stinging habit. Social insects that sting humans or other large animals do so as a defensive measure. They are either trying to protect themselves or their nest.

Stinging insects away from their nest are unlikely to be aggressive. Bees visiting a flower or wasps searching for a caterpillar are quite docile—unless they are physically restrained. Then they will use their stingers to protect themselves.

Stingers of insects are modified egg-laying devices called ovipositors. So only female insects sting. In most social insect colonies, the workers are non-reproductive females with stingers. In general, these are the social insects humans most often encounter since they, as their name suggests, do the work.

In bumble bees and honey bees, the males are called drones and can be handled without fear of being stung. Of course, you need to be able to tell the male from the female. In honey bees, the drones are larger than the workers. In bumble bees, the males have much larger eyes than their sisters. These characteristics aren't of much help, unless the males and females are side by side for observation.

In carpenter bees and cicada-killer hornets, the males pretend like they can sting. They will actually land on a person and act like they are stinging with their abdomen. It is all a show; but a show that works, because most animals aren't willing to find out if the insect is a he or she by being stung!

The goal of the stinging insect is to inject poison into the offending animal. The stinger is just the needle to do the job. Honey bee stingers are barbed. When the stinger is jabbed into the skin, the barbs—like those on a fish hook—keep the stinger from being removed. When the bee leaves, she pulls the end of her abdomen off. This results in the death of the bee. This action has been called the ultimate act of altruism—giving your life for the good of the colony. The human equivalent is a soldier's loss of life in wartime.

Stingers of bumble bees and wasps are sharp, smooth needles without barbs. Consequently, these bees and wasps can sting multiple times.

In all instances, it is best not to disturb nests of bees and wasps. They are likely to protect the colony at your expense. As Hoosier poet James Whitcomb Riley once penned in reference to a bumble bee: "If you don't think they can sting—you'll see!"

teraction with humans. Such encounters provide fodder for writers and poets. Indeed, it is a rare wordsmith that has not taken a pen in hand to wax eloquent about the virtues of ants.

Aesop used the lowly ant to extol the virtues of hard work and planning in his well-known fable, "The Ant and the Grasshopper." Who can forget the moral of that story: It is thrifty to prepare today for the wants of tomorrow.

Folklore includes many references to ants, especially to their small size: "A coconut shell full of water is an ocean to an ant," and "Even an ant is eight spans long as measured by its own hand." But folklore also speculates on the potential of ants if their size were equal to their might: "What would the ant do if she but had the head of a bull?"

But while they are small in stature, their wisdom looms large. "Ants never bend their course to an empty granary," according to folklore. And the Bible includes references to the wisdom of ants. In Proverbs 30:25 we read: "The ants are a people not strong, yet they prepare their meat in the summer." In Proverbs 6:6 we are admonished: "Go to the ant thou sluggard; consider her ways and be wise."

Odgen Nash penned a few well-chosen words about ants.

The ant has made himself illustrious,
By constant industry industrious.
So what. Would you be calm and placid,
If you were filled with formic acid?

Nash refers to formic acid, a chemical commonly found in ants that is used by many to communicate with each other. "The word goes out in formic," as penned by Robert Frost, provides another account of the chemical communication used by ants. Frost describes the regimented structure of these social insects in his aptly named poem, "Departmental."

We can all learn a few things from ants. In fact, ants are great teachers. Such is communicated by this proverb: "None teaches better than the ant, and she says nothing."

'Nuff said!

Ants in Your Pants

Ants. They're everywhere, they're everywhere! Or so it seems. Ants can be found from the Arctic regions to the tropics. From the mountain peaks to the seashore. From the kitchen counter to the backwoods picnic, they'll be there.

Ants are one of the most recognizable of insects. It is no wonder. There are over 10,000 species of ants and everyone, at some time, has encountered an ant or two.

There are black ants and red ants. There are carpenter ants and fire ants. Some ants are predators, feeding on other animals—primarily insects. Some ants are scavengers, including the fire ants. Others harvest seeds and some tend aphids and use the aphids' excreta, called honeydew, as a food resource.

Often in the spring and summer large numbers of winged ants can be seen swarming about. Winged ants are the sexual forms. They possess wings for only a short time and use them to disperse from old nests in an effort to establish new colonies.

Many of us are familiar with ants because we consider them pests—whether in the sugar bowl, lawn or field. Ants also are beneficial and have even been used in medicine. In early medical literature it is noted that black ants could be used to close incisions. This was accomplished by placing an ant so that its wide open jaws snap shut across the incision. The ant's head is then pinched from its body and the jaws remain closed until the wound heals.

Formic acid is the chemical commonly found in ants that they use in their sting and as a trail-marking substance. It is also included in the family name of ants, Formicidae.

No doubt, the potential bite and sting of ants combined with their presence in the most unlikely of places has given rise to the time-honored description of a fidgety person. Indeed everyone can imagine what it would be like to "have ants in their pants!"

Aphids and Ants: Another Dynamic Duo

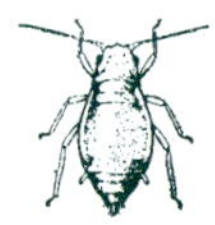

Throughout the ages, mankind has had its share of dynamic duos—Adam and Eve, Samson and Delilah, Anthony and Cleopatra, Bonnie and Clyde, and Batman and Robin to name a few.

Insects also have their duos. Ants and aphids are often found together. To many gardeners and homeowners neither ants nor aphids are favorite insects. Both can be pests, and when they show up together, some gardeners panic.

Aphids (sometimes called plant lice) and ants have a mutually beneficial relationship. The aphids provide a food resource for the ants, and the ants provide protection for the aphids. Some say that ants are farmers and aphids are their cows.

Ants feed on the honeydew produced by the aphids. Aphids feed on the sap of plants, filtering out the portions they need and excreting the rest. The excreted, sticky liquid, known as honeydew, includes a high concentration of sugar, which is highly prized by the ants.

It is the sticky honeydew that sometimes gums up automobiles parked under an aphid-infested tree. The sugar concentration of the honeydew also provides an ideal place for fungi to grow. Sooty mold growing on aphid honeydew frequently turns aphid-infested plants black. This adds an unsightly insult to gardeners' prized plants.

Since the ants are harvesting the honeydew produced by the aphids, they become quite possessive of their "cows." The ants protect the aphids from predators, like ladybird beetles and lacewings, and parasites. Ants even carry aphids to uninfested plants to provide fresh pastures for their herd.

That famous duo of song composers, Rodgers and Hammerstein, once noted that love and marriage go together like a horse and carriage—at least in Oklahoma. Maybe they also could have included aphids and ants in the song. After all, in love and marriage and between aphids and ants, sweetness is the tie that binds.

Insect Carpenters

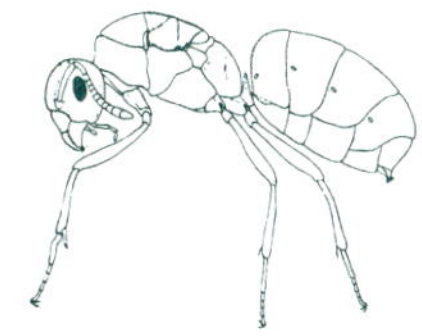

Carpenter bees and carpenter ants—these insects don't carry hammers or saws, or belong to a local carpenter's union. However, their wood work creates few friends in the human world.

Bee and ant carpenters build their homes by hollowing out wood. This is accomplished by using their strong mandibles. They literally chew out a new dwelling. In the woods this home-building activity is largely unnoticed by humans. However, when either of these insects decides to make their home in our homes, we are less than hospitable.

These insect carpenters can weaken the structure of a house by building their domiciles. So we call them pests and try to get rid of them.

Carpenter bees resemble bumble bees. However, the carpenter bee is nearly black, while the bumble bee has yellow markings. Also, bumble bees are fuzzy, while carpenter bees are slick. They excavate a nesting burrow in exposed timbers like rafters and porch railings. The entry hole to the burrow is normally upward into the wood for about an inch. The burrow then turns and runs with the grain of the wood. The female bee makes cells which are filled with pollen and nectar and separated with wood pulp. One larva lives in each cell.

Much of the concern about carpenter bees is due to the habit of the males flying around the nest area. Males cannot sting, and females, unlike bumble bees, will not defend their nest. They only will sting if mishandled.

Carpenter ants are large and black. They can be either winged or wingless. The presence of carpenter ants in a home does not mean they are nesting there. These ants will look for food away from their nest and will often search a human dwelling for something to eat.

Carpenter ants, like human carpenters, leave sawdust as evidence of their work. A pile of sawdust on the carpet may indicate that these insect carpenters have constructed their new home in yours. If this is the case, it is probably wise to do something about the ants. Their nests can weaken the structure of our homes.

Most homeowners find it necessary from time to time to have carpentry work done. But not the kind provided by carpenter bees or carpenter ants.

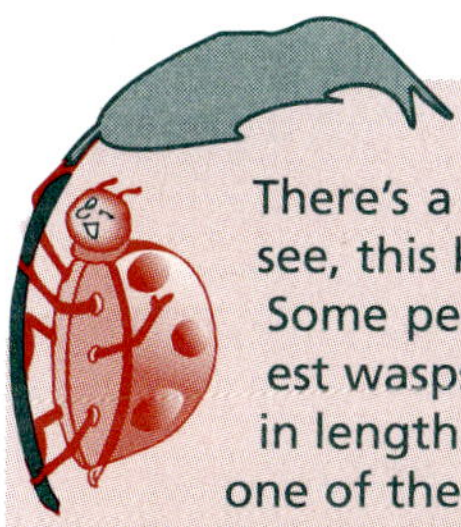

Killer on the Loose

There's a killer on the loose! But neither the FBI nor the local police departments are concerned. You see, this killer is of the insect kind. It's a cicada killer.

Some people are frightened by these wasps, quite understandably. Cicada killers are one of the largest wasps in North America. These rusty-brown insects with yellow stripes range from 1 to 1-1/2 inches in length. That size wasp, one is to assume, has a good-sized stinger! Indeed, the female is said to have one of the most severe stings of any insect.

For the most part, people's fear of cicada killers is unfounded. These wasps reserve most of their stinging activity for—you guessed it—cicadas. The adult wasps catch cicadas. Often the catch is made while both are on the wing. The cicada killer then stings its victim, a process that paralyzes the cicada. Before the cicada is paralyzed, it squeals loudly. It makes quite a sight—this B-52 of the wasp world headed toward home with a squealing cicada in its legs.

The cicada killer uses the cicada to feed its young. The wasp first digs a hole in the ground. At the end of the hole, 6 inches or more in depth, the mother-to-be places one or more cicadas. She then lays an egg, which hatches in two to three days.

The newly hatched larvae begin to feed. The adult wasp has guaranteed a fresh food supply for her young by paralyzing the cicada. Not just any cicada, mind you, but a particular species, the dog-day cicada. The cicada is immobilized, but remains alive. The larvae keep the cicada alive by selective feeding. They destroy nonessential parts of the host first, leaving the vital organs until the final stages of feeding.

By fall the larvae have completed development. They then pupate and hibernate until the following summer, when a new crop of cicada killers begin to ply their trade.

It is the wasp's behavior in the vicinity of the nesting sites that frightens people. The wasp buzzes around, hovering menacingly, as it checks out intruders. Fortunately, it seldom resorts to warfare as a means of protecting the nest.

When in the vicinity of the nesting grounds of cicada killers, the best approach is to keep your cool. And it might be prudent to minimize noises similar to those made by a cicada!

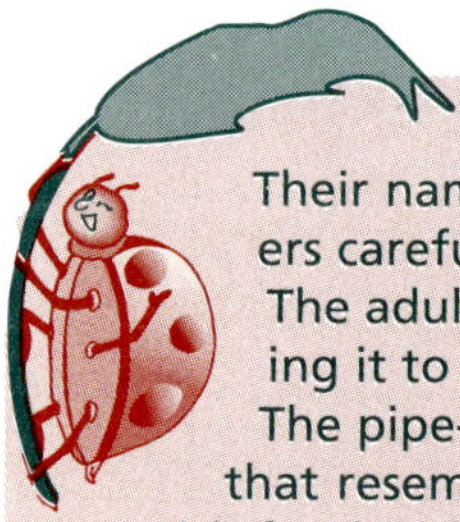

Their Names Are Mud . . . Daubers

Their names may be mud, but you won't see them slinging any. In fact, the mothers-to-be mud daubers carefully construct cozy nests for their young.

The adult female wasp frequents creek banks or puddles, gathering mud in her mouthparts and carrying it to the nest site. There she fashions the mud into a nest characteristic of her species.

The pipe-organ mud dauber forms nests in series of 3 to 4 inch tubes constructed side by side in a way that resembles the pipes of a church organ. Each pipe of the nest contains cells which are provisioned with food such as spiders and caterpillars.

For those mud daubers choosing small spiders as meals for their young, as many as 20 can be placed in a cell. Each spider is stung by the wasp and paralyzed. Once the food is in place, the wasp lays an egg and the cell is sealed.

The newly-hatched wasp larva feeds on the spiders until it is fully developed. Then the larva forms a pupal cell and changes into a wasp. The following spring, the new wasp chews a hole through the mud walls of the nest and emerges. Once out, the adult wasp begins the age-old ritual of "mud daubing" a new nest.

While mud daubers sometimes act as if they might sting humans, they seldom do. They are peaceful neighbors who build their nests in our attics, garages and barns. Mud daubers even have been known to build nests in unused items such as sleds, boots, and picnic baskets.

By the time the young emerge, the parent will be dead—a victim of old age, hard work or the first freeze. Mud daubers work very hard to provide for their offspring even though they will never live long enough to see them. For mud daubers, the term "generation gap" is appropriate.

The Socializers

1. What is a bee cleansing flight?
2. What is the larval food of bald-faced hornets?
3. Robert Frost's poem "Departmental" is about which group of insects?
4. What substance is utilized by ants for trail marking and as a sting?
5. Where would the larvae of cicada-killers be found?
6. What is another name for plant lice?
7. What insects have been said to be cows for ants?

CHAPTER 11

Bugs (Hemiptera and Homoptera)

The word "bug" has been used as a catch all for naming any kind of insect, as well as other undesirable creatures. To an entomologist, a bug has always been any insect found in the Order Hemiptera. Recent research indicates that this might not be entirely true, giving "bug" yet another meaning. Traditionally, the Orders Hemiptera and Homoptera have been classified separately. Today, there is evidence suggesting they should be combined into one Order, although we will consider them as separate Orders. Homoptera (from the Greek *homo* meaning like and *pteron* meaning wing) include such insects as cicadas, leafhoppers, treehoppers, spittlebugs, aphids, and scales. Hemiptera (from the Greek *hemi* meaning one-half and *pteron* meaning wing) include such insects as water striders, bed bugs, squash bugs, and giant water bugs. There are over 65,000 species of these insects described, with 11,000 found in North America. They range in size from minute (0.3 mm, a fine grain of sand) to quite large (100 mm, the height of a small cup of coffee).

Both Hemiptera and Homoptera have piercing-sucking mouthparts. Homoptera use these mouthparts, originating on the posterior portion, or rear, of their head, to feed on the fluids of plants. Hemiptera use these mouthparts, originating on the anterior portion, or front, of their head, to feed on the fluids of plants or animals (being either a predator or a parasite). Beware! Some of these can produce quite a painful bite! In the adult stage, these insects have two pairs of wings. Homoptera hold their wings either rooflike or flat over the body, while Hemiptera hold their wings flat over the body. The forewings of Homoptera are uniform in structure, whereas the forewings of Hemiptera are called hemelytron, being thickened at the base and membranous in the distal half. Most of these insects have a distinct triangle on their back where their wings join or cross, called a scutellum. The Hemiptera generally have long antennae, whereas the Homoptera have short antennae. Some of the Hemiptera possess a gland that produces quite a stink, and sure enough, they are called stink bugs!

All Hemiptera and Homoptera go through incomplete metamorphosis. The immature insects look just like their parents, without the size or wings. The larvae feed on the same food as the adults, and certain species are found in large numbers together.

So What's Bugging You?

Little boys have been known to "bug" older sisters and sometimes parents. Operatives of secret organizations fear "bugged" meeting rooms. Gossips have sometimes put a "bug" in a willing listener's ear. And we have all slept "snug as a bug in a rug" on a cold winter's night.

The word "bug" conjures up all kinds of ideas in people's minds. For instance, mentally ill people are sometimes incarcerated in a "bug house," and society has come to regard such folks as "buggy." Most of us have at some time in our lives "committed a bug-a-boo," to the dismay of our colleagues. But what do children look like when they are "cute as a bug's ear?"

We may never know the answer to that question, since bugs don't have real ears, but most of our current uses of the word "bug" reflect an ancient meaning similar to the Celtic word "bwg," (pronounced BOOG), which meant ghost or spirit. Such an idea, and the word, is incorporated into the thought of the bogeyman—that mystical spirit of darkness sometimes used by parents to keep children in line.

The word "bwg" was probably first used to describe the bed bug. This pest of humans is rather reclusive in its habits. It hides in cracks and crevices of houses during the day. It emerges at night and, under the cover of darkness, attacks sleeping humans. To the ancient Celts who fell victim to this insect, it must surely have seemed that they were victims of a ghost or spirit, thus, the name bwg. We call the shy creature a bed bug and, to this day, admonish, "Good night, sleep tight, and don't let the bed bugs bite."

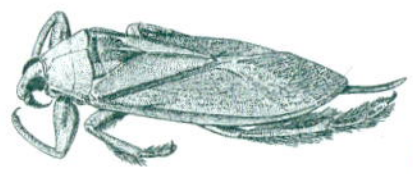

Scientifically, bug is used to describe members of the insect order Hemiptera, the true bugs of the insect world. Bugs include such aquatic insects as the giant water bug, water striders and backswimmers. And on our plants, we sometimes find squash bugs, stink bugs, and boxelder bugs. Ambush bugs and assassin bugs are, as their names suggest, predators—mostly on other insects. Kissing bugs, like the bed bugs, are pests of humans. They get their name from the habit of biting folks around the mouth.

Regardless of its scientific merit, the word "bug" is frequently used in reference to any insect—an entomological error! You see, all bugs are insects, but not all insects are bugs.

Such a mistake was made by the computer folks who discovered that an insect had short-circuited their computer. The insect was a moth—not a bug—but after its removal, they coined the term "debug" to describe the process of fixing a nonfunctioning computer.

The Treetops Are Alive with the Sounds of Cicadas

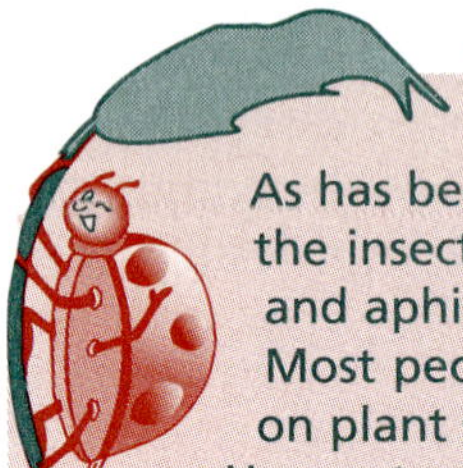

As has been the case for many millennia, summer brings the sound of cicadas. Cicadas are members of the insect order Homoptera. This order includes plant hoppers, leaf hoppers, whiteflies, scale insects and aphids.

Most people recognize these insects as pests. And many Homoptera do damage to plants by feeding on plant sap and sometimes transmitting plant diseases.

Homoptera generally are small insects. Cicadas are an exception. They range in size from one to two inches in length.

Cicadas are well known because of their sound production. Two organs on the abdomen are used to create sounds. Each organ includes a membrane and a ribbed structure called a tymbal. When the tymbal is vibrated by strong muscles, the "song" is produced and amplified by the membrane.

Cicada songs are produced only by males. The male sings to attract the female of the species. Each type of cicada has a specific song, but to humans, most of the songs can be described as a buzz. When many cicadas are singing in unison, the buzz can become quite annoying—and loud! So loud that some folks have described the music as a deafening racket.

Regardless of the human perception of the quality of the sound of cicadas, their singing certainly attracts attention. Some authorities hold that the grasshopper in Aesop's Fable of the grasshopper and the ant actually may have been a cicada in the original tale. Both insects were well known to ancient peoples. Both grasshoppers and cicadas emerge in large numbers and produce sound. And essential to the fable, neither insect plans for the winter at least by laying in food as do the ants.

Confusion between cicadas and grasshoppers has occurred several times throughout recorded history. Such confusion occurred in North America when the first European settlers arrived. They encountered great emergences of cicadas and called them locusts, the name commonly used in other parts of the world for migratory grasshoppers. As a result, in the United States, we frequently refer to cicadas as locusts.

By either name, locust or cicada, the insect fascinates us as much as it did people in Aesop's day. Biologically, the life cycle is interesting. The adult female lays eggs in slits under the bark of the twigs of trees. The eggs hatch and the young cicada drops to the ground where it crawls into the soil. In the soil, it feeds on the roots of trees.

After its immature stage is complete, the cicada crawls out of the soil, mostly during nighttime hours. From there, it climbs up tree trunks, sides of houses or fence posts where it attaches and emerges from its immature skin. The skin, called a shell by many people, is left for children to collect.

Cicadas, depending on species, spend anywhere from 1 to 17 years underground during the immature stage. The species with the 17-year life cycle are known as periodical cicadas. There are also periodical cicadas with a 13-year life cycle. Either way, the periodical cicadas have the longest life cycle of any insect, except the termite queens of Africa.

The emergences of periodical cicadas sometimes results in millions of the adults present at the same time. Under these circumstances, the trees are literally alive with flying and singing insects. Native Americans were well aware of the singing cicadas. But these resourceful people enjoyed more than the song of the insect. Many Native Americans collected cicadas as they emerged from the soil, roasted them over a fire and ate them. For them, it was a delicacy complete with dinner music, provided by the delicacy!

Oh well, most entomologists don't worry about the misuse of the term "bug."

Leafhoppers by the Zillions

Leafhoppers seem to be everywhere during the summer. These small insects, measuring only millimeters in length, can be found on almost all types of plants.

They feed by taking sap from leaves. Their feeding pattern plus their habit of jumping when disturbed has given them the name "leafhoppers."

Many leafhoppers have a single generation per year and spend the winter in the adult or the egg stage. Some are seasonal invaders, coming into the midwestern states in the spring as passengers on a spring weather system.

Leafhoppers are important pests of plants. Damage is caused in several ways, including stress due to removing sap. Some species injure plants by laying eggs in twigs. Other species are vectors of important plant diseases, including potato yellow dwarf and corn stunt.

One of the most common pests, of this group of insects, is the potato leafhopper. Though it is a pest of potatoes, it also feeds on alfalfa and soybeans. When the potato leafhopper attacks alfalfa, its saliva produces a toxic effect, causing the leaves to turn yellow in a V-shaped pattern.

You don't have to grow potatoes or alfalfa to consider the potato leafhopper a pest. This insect invades our homes much too frequently. The potato leafhopper really doesn't want to be a houseguest. It just happens to be an insect that cannot stay away from lighted windows.

The potato leafhopper is so small that it can go through window screens in search of the nearest light fixture. By the next morning, leafhopper carcasses are strewn like fallen infantrymen across counter tops, in light fixtures and on window sills.

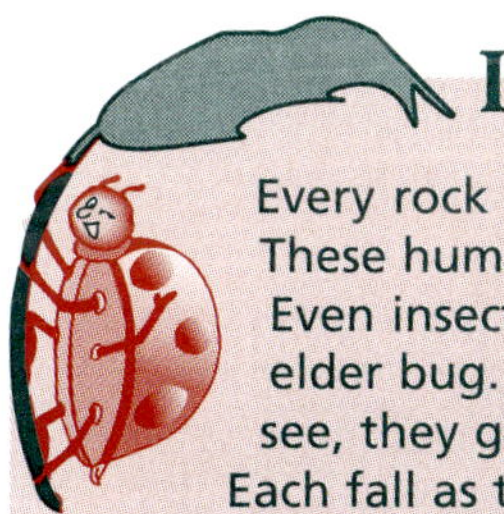

Insect Groupies

Every rock and movie star has groupies. So do athletes. On the golf links Arnie Palmer had his army. These human hangers-on are common around the rich and famous.

Even insects have groupies—at least they hang around in groups. One such insect groupie is the boxelder bug. No famous individual attracts these bugs. They just hang around for the warmth of it. You see, they gather in groups to absorb the rays of the sun.

Each fall as the season turns cooler, boxelder bugs begin to congregate in great numbers. These groups of strikingly marked red and black insects can be found everywhere—on trunks of trees, on porches, on walls and even on sidewalks. Often the presence of large numbers of boxelder bugs causes a great deal of concern among humans in the vicinity.

Such concern is unwarranted. Boxelder bugs can neither sting nor bite. They do not harm food or clothing. However, they do seek protection during the winter months in sheltered places. Sometimes these places are our homes. However, most homeowners take a fairly dim view of sharing their domicile with a bunch of insect groupies.

Because of such concern, homeowners have been known to take drastic measures to eliminate boxelder bugs. One such method is to attempt to burn the bugs. This approach generally involves a homemade torch on a pole. But there are secondary effects of such an effort. For instance, singed hair—from falling pieces of disintegrating torch. And we cannot forget the possibility of an emergency run by the fire department if the fire gets a little out of hand!

A better approach is to spray a common insecticide. Another way to reduce the problem is to eliminate boxelder trees in the vicinity of the house. These insects feed primarily on the seeds of the boxelder tree, but they also consume seeds of ash trees. Removal of the food source—in this case, the trees—will reduce the population to low levels.

Boxelder bugs characteristically have a disagreeable odor and taste. This is well known to people who have accidentally encountered this insect. Such a bad taste—a fact advertised by its bright color—keeps the insect from being eaten by insect predators.

Boxelder bugs aren't called boxelder bugs by everyone. In parts of the Midwest, these insects are known as Democrats, which is an unusual name compared to other common names of insects. I was once told that the name Democrat reflected the habits of these insects—because they always hang around in little groups and raise a stink. Of course, the fellow who mentioned that fact was a Republican!

Even insects like the boxelder bug aren't immune to politics.

Home owners who have enlisted bug zappers in their personal battle-of-the-bugs will find that their prized device has claimed leafhoppers by the bucketful. Still, many leafhoppers manage to find their way into the house. In the battle with insects, we humans have come to realize that numbers can sometimes overcome the most modern devices.

Plant Lice

Springtime—that time of the year when plant fanciers' thoughts turn to green growing things—and to insects! Indeed, each spring, growers of ornamental and food plants prepare to assist their leafy-green wards in the "battle of the bugs."

About half of all insects feed on plants. It's a natural ecological process; the plant is the producer, and the insect, a consumer. Most plant owners, however, take a dim view of such insect activity.

Almost all plants are associated with one or more species of aphids. Sometimes called plant lice, aphids are small insects about the size of the head of a pin. Some have wings and some are wingless, even within the same species.

Aphids are one of the suckers of the insect world. They get their food by sucking sap from the leaves and stems of plants. In doing so, they can weaken the plant, cause abnormal tissue growth, and even transmit disease organisms. That's enough to send even the most seasoned plant grower looking for the spray can.

How can a little insect provoke such fear? It has to do with numbers. You see, aphid populations can grow very rapidly. As Erasmus Darwin, grandfather of Charles, once wrote:

The countless aphides, prolific tribe,
with greedy trunks the honey's sap imbibe;
swarm on each leaf with eggs or embryons big,
and pendant nations tenant every twig.

Many aphids give birth to live young. Some of these young can themselves become mothers in about a week. To help speed the process, some species have cut out the middle man—they reproduce without mating. No need to slow down the process by taking time to find a mate.

Aphids aren't despised by all creatures. For example, ants use the aphids' waste material, called honeydew because it is sweet, as a food resource. So attached are some ants to honeydew that they protect the aphids that produce it. This has prompted some scientists to call aphids "ants' cows." Ants are even known to move aphids from one plant to another as they manage their herd.

A few insects see the aphids themselves as the meal. Ladybird beetles are famous as predators on aphids. So are lacewings. Because the predators eat so many aphids, gardeners love these beneficial insects. Some people even purchase ladybird beetles for release to help in aphid control. However, it is important to make sure that aphids are present before turning the ladybird beetles loose. If aphids aren't present at the time of release, the ladybird beetles won't have food to eat and will leave.

In addition to causing plant stress, aphids sometimes create other problems by their presence. As aphids feed on trees, their honeydew can drift downward and leave rather unsightly spots on automobiles. Honeydew on plants provides a great place for sooty mold to grow, which results in a black layer on the surface of the plant. It is not a pretty sight to the plant owner.

All of this from a little insect called an aphid. It's no wonder that they are called plant lice. The very thought of having them makes most gardeners get itchy!

The Year of the Cicada

The cicadas are coming!
The cicadas are coming!

Years ago when much of the eastern United States was covered with hardwood forests, a brood of periodical cicadas probably emerged each year. Today, some broods have been lost entirely, and most are confined to localized areas.

Periodical cicadas should not be confused with annual cicadas, which emerge during July and August. Annual cicadas are sometimes called "harvestmen" or "dog-day cicadas" because of their emergence time, which is associated with the dog days of summer.

The periodical cicada has the longest developmental period of any insect. As the name suggests, it takes 17 years for the insect to complete its life cycle.

Periodical cicadas begin their life as an egg laid under the bark of woody plants. The eggs hatch and the nymphs drop to the ground, where they dig into the soil and feed by sucking sap from the roots of trees. Seventeen years later, the nymphs emerge

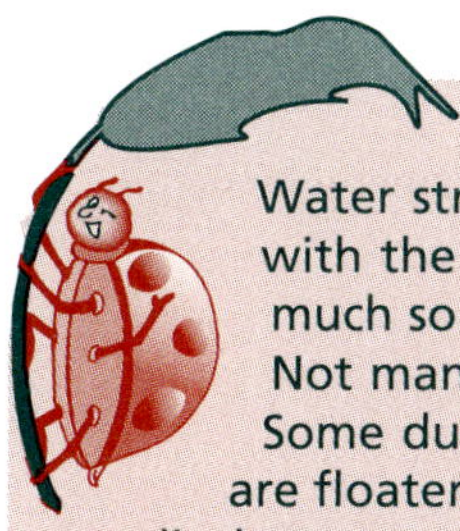

Water Striders Use Tension to Stay Afloat

Water striders are appropriately named. These insects are able to glide over the surface of the water with the greatest of ease. Being able to walk on water is a miracle of sorts in the animal world—so much so that some people actually refer to water striders as "Jesus Bugs."

Not many creatures in the animal world actually live on the water surface as do these insect striders. Some ducks and geese spend much of their life on water; so do some seagulls and terns. These birds are floaters. They are animal boats. Waterproof feathers form the bird equivalent of a hull. The creature displaces enough water to keep it on the surface.

Water striders, though, are different. They manage to stay on the surface by using the surface tension of the water to hold them up. Remember in grammar school how, with great care, students were able to float a needle on the surface of the water in a glass? The secret was to not break the surface tension of the water. Otherwise, the needle did a miniature version of the Titanic and headed to the bottom of the glass.

So how does this insect strider manage to pull off this miracle? It has the right stuff. Its insect equivalent of toes, called tarsae, are covered with very fine hairs that are difficult to wet. These nonwetting hairs do not break the tension of the water, so the insect floats, or skates, on the surface.

Those who have observed water striders closely notice that the insect appears to have only four legs. All adult insects have six legs, and the water strider is no different. However, this insect uses its front legs, which are smaller than the four used as pontoons, for capturing prey in much the same way that praying mantids use their front legs. The front legs do not contact the water, so only four legs are obvious to the casual observer.

Water striders float over the surface of the water in search of their food, which is small insects. Any hapless insect that falls into the water and stays afloat for a moment is potential food for these predators.

In general, water striders are found in protected areas of freshwater ponds or streams where the water is quiet. However, they can be seen riding the waves in fast-moving streams as well. One species lives on the surface of the ocean and can be found many miles from shore. It is one of the few insects that has any association with ocean water.

These insects lay their eggs on floating plant material. The newly hatched young skate over the water like their parents. Neither young nor adults can live under water, and if they sink they drown. Occasionally this happens. When the tiny hairs on their tarsae become wet, the insect cannot float and must climb up off the water to dry out.

Winged and wingless individuals occur in most species of water striders. It is the winged individuals that are found in unexpected places at times, for instance, on the birdbath.

What happens to water striders when the water where they are living dries up? Well, they might fly away or, in the case of a larger spring or pond, they just crawl under a rock or in the mud and hibernate until the water returns. After all, it's hard to stride on the water if there is none!

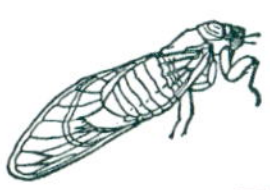

from the soil, where they shed their last shell to become winged adults.

Emergence normally occurs during nighttime hours. The cover of darkness helps protect the insect because the newly emerged adult is soft-bodied and unable to fly. At this time, it is easy prey to cicada eaters, of which there are many.

The new adults begin their appointed role in life, that of mating and laying eggs. Male cicadas attract mates by singing. The noise—or, rather, singing—is produced by two, drumlike membranes on the first segment of the abdomen.

Is it really necessary for the cicadas to make such a racket? No one knows for sure. But wouldn't most folks feel like singing at the top of their lungs after spending 17 years in underground darkness?

Bugs

1. What was the original meaning of the Celtic word "bwg"?
2. What insect is responsible for the idea of "debugging a computer?"
3. In what form do boxelder bugs overwinter?
4. What is the scientific name of the only insect order that can correctly be called "bug"?

CHAPTER 12

Other Insect Orders

Aquatic Life

If you look carefully at a rock or stick removed from a creek, you may be surprised at how many insects can be found! The entomological life that lives in freshwater has had to adapt to an environment that offers many challenges. How do these creatures breathe? Some insects remove dissolved oxygen directly from the water. A few of the insects that get oxygen this way are represented by dragonflies/ damselflies (Odonata), mayflies (Ephemeroptera), caddisflies (Trichoptera) and stoneflies (Plecoptera). Other insects break through the surface of the water to obtain oxygen directly from the air. Many of these carry a bubble with them while under water, and return to the surface to "refuel" when necessary. A few of the insects that get oxygen this way include beetles (Coleoptera), true bugs (Hemiptera), and flies (Diptera).

Living in fast moving waters can also provide quite a challenge. How do insects keep themselves from being washed away? Many aquatic insects have large claws at the end of their legs. This allows them to hold on to plants, rocks, and even sand. Insects may also have a flattened, streamlined body. This, in effect, allows water to pass right over them.

And what about food? Aquatic insects are no different than terrestrial insects when it comes to eating. Some are predators, eating other insects, including each other, small amphibians, and even fish. Others are herbivores. Some eat leafy plants while others dine on a variety of algae. Still others are omnivores, dining on both plants and animals.

Another challenge faced by aquatic insects is reproduction. Where do aquatic insects mate? Many of the insects found under water are in the larval stage of life. When these larvae mature, they metamorphose into adults and leave the aquatic environment. Some of these insects do not feed as adults and live for a very short amount of time (mayflies), while others can have a long adult life (dragonflies). Virtually all of these insects will return to the water to lay their eggs. The world of aquatic insects remains a mystery to many, yet they are a diverse and fascinating group. The next time you're in the water, see what you can find.

Insect Dragons

Skimmers, Biddies, and Darners—no, these aren't the latest rock bands. They are dragonflies.

Dragonflies are among the oldest of insects. Some call them living fossils. Today's dragonflies look about like their ancestors that lived nearly 300 million years ago. Dragonflies roamed the skies during the 100-million-year reign of the dinosaurs. One such ancient dragonfly was itself dinosaur-sized. This insect had a wingspan of nearly 27 inches.

Immature dragonflies, called nymphs, live in the water. Adults do fly far from water on their aerial forays, but they are most commonly found near water habitats.

Both adults and nymphs are predators. They feed primarily on other insects, although large dragonfly nymphs have been known to capture and devour small fish.

Adult dragonflies capture prey while on the wing. Although the dragonfly has a fearsome-looking set of jaws, it procures its meals by an unusual method. The dragonfly shapes its six legs into a basket. Then, in an insect version of an aerial dogfight, overtakes prey from behind, snaring the potential meal in its legs.

The dragonfly sometimes eats on the wing—sort of a "meal to go" in the insect world. For a more leisurely dining experience, the dragonfly will perch on vegetation before devouring the hapless victim.

Because dragonflies destroy a number of insects, including mosquitoes, gnats and flies, they are beneficial. Even though dragonflies are predators, they are harmless to people. These insects neither sting nor bite. However, dragonflies have not always been considered harmless.

There are a number of superstitions associated with the dragonflies. For instance, in some rural areas of the United States, the common name "snake doctor" was acquired because people assumed that the dragonfly acted as a guard to snakes. This may have originated because both snakes and dragonflies share marshy habitats.

Another superstition held that dragonflies could sew together various parts of the human body, including lips, nostrils, eyelids and ears. This belief gave rise to the common name darning needle or devil's darning needle.

Some of the fear of dragonflies, and some of the superstitions, are probably based on the fearsome look of the insect. However, part of the mystique of the dragonfly is due to its territorial nature. Male dragonflies set up territories just as do certain birds and mammals. The dragonfly then defends its territory against other dragonflies. Such territorial defense leads to spectacular battles among these master aerialists. The dragonflies also check out other intruders in their territory—a surveillance activity that some folks might interpret as hostile behavior.

The next time you spy a beautiful, brightly colored dragonfly doing aerial maneuvers, pause for a moment and watch closely. You are watching an animal that has been flying for 300 million years. With all of that practice, it's no wonder dragonflies are good at flying!

Winter Stoneflies

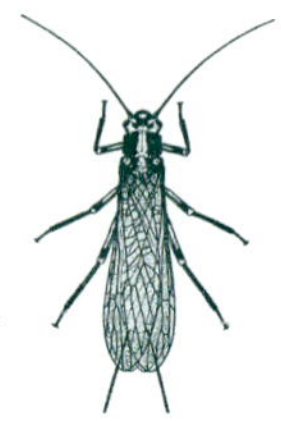

Insects, like snakes, frogs and salamanders are cold-blooded organisms. These animals cannot regulate their body temperature and are generally inactive during the winter.

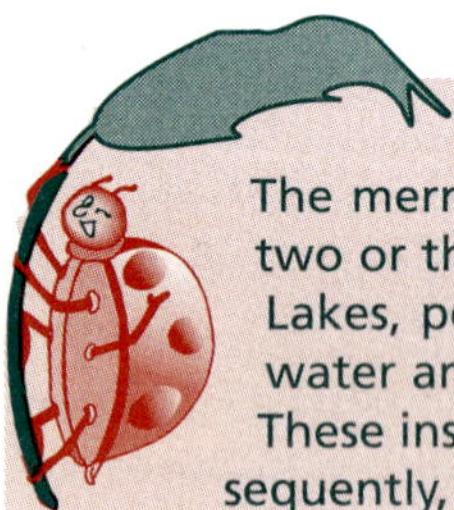

Fleeting Flies of May

The merry month of May just wouldn't be complete without mayflies. These soft-bodied insects with two or three tails abound this month.

Lakes, ponds and streams are the playgrounds of the mayflies, for their immature stages live in the water and feed upon algae and other aquatic plants.

These insects in both the adult and immature stages are also an important source of food for fish. Consequently, fishermen use artificial lures modeled after the mayfly in their attempts to catch the big ones.

The aquatic stages of mayflies can last a year or more. The adult has one of the shortest life spans of any insect. In fact, these insects are classified in the order Ephemeroptera which is based on the Greek word "ephemero," meaning for a day. This refers to the adults which will live for as little as 24 hours. The adults have reduced mouthparts and do not feed.

The short time the adults have on the wing is devoted to their single biological function, reproduction. Mayflies often emerge in enormous numbers and form swarms. These swarms are normally composed of males which fly up and down in unison in the mayfly version of a mating dance. When a female flies into the swarm, she is claimed by a male and spirited away on a mating flight. The eggs are laid on the surface of the water. Once the biological function of the adults has been completed they die. The dead and dying mayflies don't go to waste; they provide an all-you-can-eat smorgasbord for the fish.

Houses and towns in areas where large emergences of mayflies occur are frequently inundated with these insects. They are attracted to lights where many die leaving piles of dead mayflies several inches deep under store windows or porch lights. In years of high emergence, mayfly bodies may accumulate on river bridges, causing the surface to become slick and hazardous. In some instances, snowplows have been called out to clear the bridges of dead mayflies.

Paul Fleischman in his award-winning book *Joyful Noise* begins his poem "Mayflies" with these lines:

Your moment—Mayfly month
Your hour—Mayfly year
Your trifling day—Our life

Mayfly adults are indeed ephemeral, that's why entomologists classify them as Ephemeroptera. Such is the fleeting life of the mayfly adult. One last frantic day to ensure the next generation.

Some insects, such as the familiar cockroach, cope with cold temperatures by living in our dwellings. But a few hardy insect species are seen during the winter months in the great out-of-doors.

One such insect, the winter stonefly, is appropriately named—at least the season in its name is accurate. Winter stoneflies can be found crawling around during sunny days from January through April.

But stoneflies really aren't flies at all. Some do fly, although their aerial forays are less than spectacular. A few species of stoneflies have given up flying altogether, and for good reason: they lack wings. These wingless individuals depend on their six little feet to get them where they want to go.

Technically, stoneflies are members of the insect order Plecoptera. Insects in this order are rather soft-bodied and have antennae-like appendages, called cerci, protruding from their abdomen.

These rather drab-colored insects are found near streams or rocky lake shores. The immature stoneflies, called nymphs, live in the water and are often found under stones. That's the basis for their common name, stonefly.

The stoneflies that show up during the winter months are normally active in air pockets that form between the ice and the surface of the water of streams. At a time of year when most insects are out of sight, stoneflies take advantage of the higher temperatures in such miniature greenhouses to be active.

During warm seasons when the ice disappears from streams in midwinter, stoneflies show up in abundance on fence posts, bridges and other structures in the vicinity of the streams where they emerged.

Why are these insects active during the winter? No one knows for sure. One suggestion is that it avoids competition with other insects. The stonefly has discovered a time of year when it doesn't have to compete with the teeming hordes of insects that show up during the warmer months. As far as insects go, it has the whole world to itself.

So what does the winter stonefly do under such circumstances? Like any good adult insect, it does what nature has designed it to do: it mates and lays eggs.

And what does it eat? Nothing at all. You see, the winter stonefly doesn't feed in the adult stage, which is fortunate, since there is probably very little to eat under the ice on a winter's day!

La Cucaracha, La Cucaracha

La cucaracha, the cockroach! Probably no insect conjures up more unpleasant thoughts in human minds than the cockroach.

Cockroaches are a very old type of insect. Some scientists estimate that roaches have been on earth for over 300 million years. During most of that time, roaches have devoted their considerable energies to breaking down dead plant material. From an ecological perspective, that is a very noble thing to do.

However, at some point in history, some of these six-legged scavengers took up abode in our dwellings. That, in the humble opinion of humans, was a downright unfriendly thing to do. So we devote a lot of time and energy trying to keep roaches from our homes.

There is nothing inherently bad about sharing our domicile with roaches. To be sure, they have been implicated in disease transmission. And, without question, they are common in unsanitary conditions. But mostly it is just plain socially disgraceful to share quarters with them.

To the modern housekeeper, having roaches is akin to being placed in the public stocks on the town square. For that reason, we have developed ways to overcome the shock of finding roaches in our kitchens. We call some roaches "waterbugs," because somehow it is easier to report to neighbors that the exterminator only found waterbugs. And some pest control companies will make housecalls in unmarked automobiles. That, of course, is so the neighborhood will be unaware of the nature of the visit.

Cockroaches are everywhere in the world. Many have expanded their range to include foreign soil in addition to their ancestral homelands. But their names still betray their nationalities, for example, the German, Oriental, Madagascar and American roaches.

Some prefer to earn their fame closer to home though. Back in 1916, columnist Don Marquis introduced a column called "Sun Dial" in the *New York Sun* newspaper. A lowly cockroach and self-proclaimed poet, archy, invaded Marquis's office every night to write this free-verse column. A dedicated writer, archy accomplished this feat by diving head first one by one onto the typewriter keys, which explains why he never bothered to add punctuation or capital letters to his poetry. Along with

ARCHY: THE FAMOUS COCKROACH

While archy the cockroach may be famous, most people regard cockroaches as infamous. In fact, cockroaches are despised and with good reason. These pesty insects have invaded our abodes for thousands of years in spite of our best efforts to the contrary.

There is, however, one cockroach that made a name for himself. In fact this cockroach was a writer, and his name was "archy." As the story goes, archy lived in the office of *New York Sun Times* newspaper writer Don Marquis. Each evening archy would creep on all six legs to the office typewriter and hammer out a column.

Typing was not easy and archy had to use his head, "literally," for the task. He would crawl up on the carriage and do the insect equivalent of a swan dive, striking the selected key head first. This laborious approach to typing produced free verse poetry uncluttered by capitalization or punctuation since archy couldn't depress the shift key while landing on the letter key.

Don Marquis, the writer, began in 1916 to use the literary efforts of archy in a column called "Sun Dial." Even though the style is unorthodox, the wit, wisdom and commentary of archy is timeless.

From the viewpoint of a lowly cockroach, archy has a lot to say about life in general. Consider the following maxims of archy.

don't cuss the climate
it probably doesn't like you
any better
than you like it

many a man spanks his
children for
things his own
father should have
spanked out of him

live so that you
can stick out your tongue
at the insurance
doctor

insects have
their own point
of view about
civilization a man
thinks he amounts
to a great deal
but to a
flea or a
mosquito a
human being is
merely something
good to eat

Many of us probably hate to admit it, but we can learn a lot from nature, even if the teacher happens to be a cockroach.

---

his friend, mehitabel, the cat, archy provided a delightful commentary on life during those times.

In spite of our disdain for cockroaches, the wisdom of archy is worth repeating. Recognizing that cockroaches were not the favorite insects of people, archy once rejoiced:

there is always
something to be thankful
for you would not
think that a cockroach
had much ground
for optimism
but as the fishing season
opens up i grow
more and more
cheerful at the thought
that nobody ever got
the notion of using
cockroaches for bait.

. . . archy

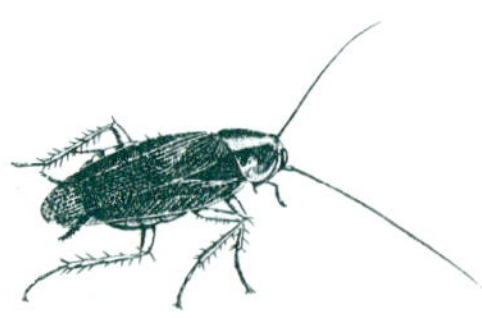

In spite of our disdain for cockroaches, archy and his descendants need not worry. They will be on this earth long after we are gone.

Arsenic and Old Lacewings

Lacewings are the "arsenic and old lace" of the insect world. These insects are small—less than a half-inch in length. To us, they appear quite fragile, soft-bodied with four clear, membranous wings. The many cross-veins in their wings create an image, well, of lace. That is the basis for their name. But to their prey, lacewings are as deadly as arsenic.

Lacewings, which are classified in the insect order Neuroptera, are considered ancient insects. Bearing resemblance to miniature prehistoric monsters, most are greenish or brown with golden-colored eyes. Their long antennae and fearsome chewing mouthparts suggest that lacewings are not part of the gentle, nectar-sipping set of the insect upper crust. Quite the opposite.

Both lacewing adults and larvae are random predators. They roam plants with the fervor of hungry diners scanning a menu as they search for insects. Aphids are a favorite food. Lacewing larvae have a pair of sickle-shaped jaws that are piercing-sucking tubes. When a prowling larva, appropriately called an aphislion, encounters an aphid, the end is swift. The victim is impaled on the lacewing's sharp jaws while fluids are sucked from the victim's body like soda through a straw.

Larval lacewings are such dedicated predators that even their brothers and sisters are potential meals. Among lacewings, the first hatched has an advantage. Newly hatched larvae begin a fervent search for food. Almost anything will do—even an unhatched lacewing egg. No amount of parental cajoling would likely prevent this ultimate expression of sibling rivalry. So lacewing mothers-to-be place their eggs on a stalk, just out of range of hungry little jaws. Of course, when larvae hatch and descend from the safe hatching perch, they're on their own in the dog-eat-dog world of lacewings and aphids.

Many gardeners have gained a great appreciation for the appetite of lacewings. In fact, lacewings have biological control potential and have even gained commercial status. Gardeners can purchase lacewings to help control the undesirable creatures that show up in their gardens—at least undesirable creatures of the insect variety. Lacewings are of no help whatsoever when it comes to hungry neighbors who seem to show up at the garden when the work is done and the harvest begins!

Termites: Nature's Wood Destruction Crew

In nature's grand scheme of things, insects have many jobs; for example, clean-up detail. Being on the clean-up crew may not be your idea of a good time, but it's a job that has to be done. Can you imagine what the world would be like if dead plants and animals remained on the surface of the earth? Combine this with the paper and beverage containers that we humans throw on the roadsides and you've got a real mess. Therefore, be thankful that clean-up is a way of life for some insects!

Termites are one of the insect cleaner-uppers. Their job is to help rid the world of dead trees. Termites destroy wood by eating it. However, eating wood is one thing, digesting it is another. To aid in wood digestion, termites have in their guts protozoa, which are microorganisms that break down cellulose.

All of this is a nice plan. However, when humans started building structures out of wood, the termites continued to do their ecological task. To a termite, dead wood is dead wood! It matters not that the wood is part of the finest house on the best street in town. Of course, when termites attack houses, people become hostile and call the little creatures pests and try and get rid of them.

Ogden Nash recognized the economic importance of termites as structural pests when he penned the following:

Some primal termites knocked on wood
And tasted it, and found it good,
And that is why your cousin May
Fell through the parlor floor today.

To many home owners the thought of termites is a sober one indeed. In the springtime, at least in the case of termites, what has been out of sight comes to mind. It is in the spring that termites swarm. Swarming is a process where termites leave the

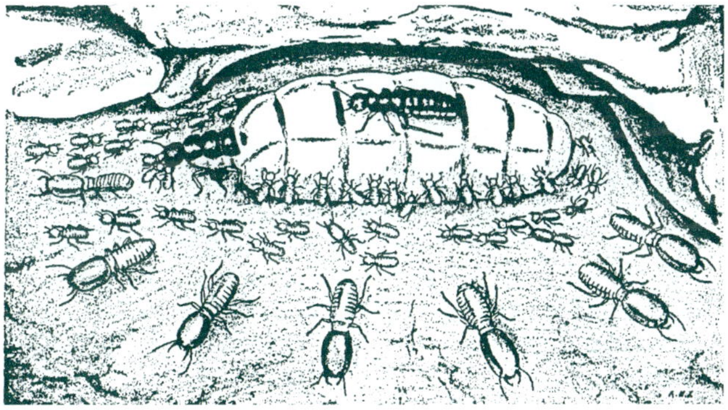

FIGURE 12.1. *Termites at Work.*

underground nest chambers and disperse. This is one time in the year when the termites can be readily seen. But they can be easily confused with ants, which also swarm.

If termites are detected in a home, the best approach for control is to select a termite control service. However, do not panic! Termites work slowly, so you have time to purchase the service at your convenience. Also, choose a reliable firm; the termite business is one that is attractive to fly-by-night businesses.

Based on fossil records, termites are one of the oldest insects. So when termites eat the wood from which we have fashioned our houses, they are doing what comes naturally, as they have for the last 350 million years or so!

Doodlebugs' Charm

James Whitcomb Riley, a Hoosier poet, recognized a doodlebug when he saw one. In his poem *The Doodlebugs' Charm* Riley revealed that it was his Uncle Sidney who introduced him to the unusual six-legged creature. What was the fascination?

Doodlebugs are strange-looking insects with large, flat heads and long, protruding sickle-like jaws. This group of insects is classified in the order of Neuroptera which includes the familiar lacewings. Both lacewings and doodlebugs are predators on other insects.

Doodlebugs live in pits constructed of dry sand or dust. These pits can sometimes measure up to two inches wide at the soil surface. During construction the doodlebugs cleverly set traps by piling

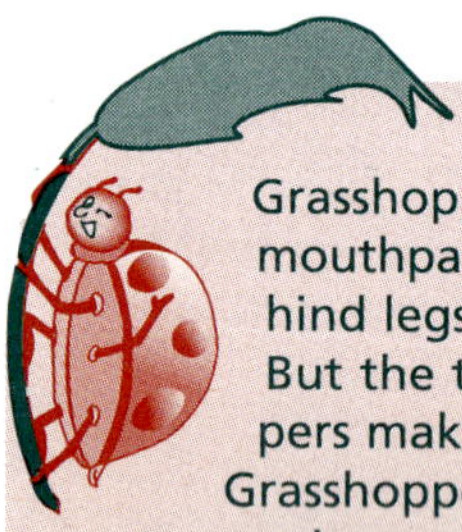

Chewing Machines

Grasshoppers are among the most recognizable insects on earth. Perhaps it's because of their chewing mouthparts. Grasshoppers have the type that show up in horror movies. Or it might be because of their hind legs. Those legs are used for jumping and, hence, the name—hopper.

But the truth is that grasshoppers are well known for their appetites. The eating habits of grasshoppers make them one of the world's worst insect pests.

Grasshoppers have been known as a pest for thousands of years. The Bible contains several references to grasshoppers—sometimes called locusts—as pests. For example, in Joel 2:3 we read, ". . . the land is as the Garden of Eden before them and behind them a desolate wilderness; yea, and nothing shall escape them." The eighth of the plagues visited upon Egypt preceding the Exodus was a great mass of locusts which, according to biblical account, ate all the vegetation on the land so that nothing was left.

Grasshoppers primarily feed on green foliage, but when food is scarce they will feed on almost anything. Tales have been told of grasshoppers eating straw hats left in the field. In the Dust Bowl days, wooden pitchfork handles were said to have been nearly consumed by hungry grasshoppers.

Grasshopper problems are most pronounced when the weather is dry. Biological control of this insect is reduced under dry conditions. For instance, under wet spring conditions, the egg masses, which are the overwintering state of grasshoppers, are attacked by a fungus disease. This disease eliminates many hoppers before they hatch.

In the arid regions of Africa, the migratory locust is a continual problem. During the drought periods of the 1930s and the 1950s, grasshopper damage to U.S. crops was high. And in unusually dry years, such as 1988, grasshopper problems normally increase.

Many approaches to grasshopper control have been used over the years. During the 1930s in the Midwest, hopperdozers were used. Hopperdozers were grasshopper-catching machines placed on the front end of a truck or tractor. As many as eight bushels of hoppers were collected per acre using one of these devices. Such a grasshopper population has been estimated to consume a ton of alfalfa hay per day in a 40-acre field.

More recently, people in Africa have tried to keep the insects from damaging crops by driving them away. Today, large swarms of grasshoppers are sometimes sprayed during their flights with an insecticide carried by aircraft.

Another way to kill grasshoppers is to include a poison in a bait. An enterprising entomologist during the 1930s even mixed poison with horse manure for use in controlling grasshoppers. Reports are that the bait worked quite well.

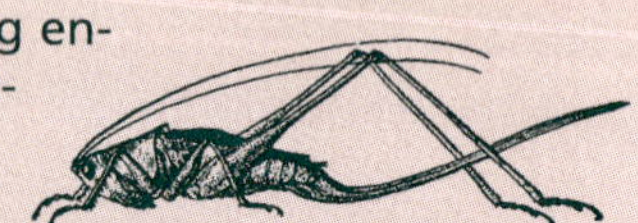

As a last resort, some peoples of the world have taken to eating grasshoppers. However, such an approach has not been widely accepted in the United States. It seems that eating up an insect problem—especially grasshoppers—just doesn't appeal to our tastes.

sand around the pits. Once inside their homes, the doodlebugs eagerly wait for their meals to "drop in." One by one insects traveling too close to the pits tumble to the doodlebug's dinner tables. Ants are frequently their victims, giving rise to another common name for doodlebugs, "antlions."

Doodlebugs are more common in the south and southwest of the country. Their pits frequently are found in groups and in very dry conditions such as under buildings.

"To call 'em up," Uncle Sidney said, "'Doodle! Doodle! Doodle-Bugs!' An' they'd poke out their head—"'Doodle-Bugs! Doodle-Bugs! Come up an' git some bread!'"

With a little strategy and a whole lot of patience, you may be able to entice the doodlebugs to show themselves. Instead of calling out their name, slowly push grains of sand into one of their pits. The curious and ravenous doodlebugs will be tempted to investigate what they think is their next meal.

If you encounter a group of the hungry larvae, then perhaps like James Whitcomb Riley, you too, will discover the doodlebugs' charm.

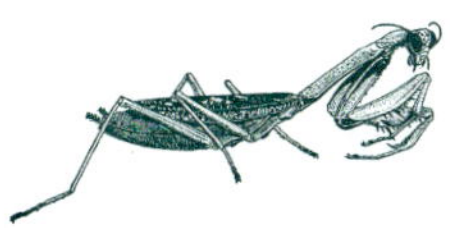

Prayerful Insects

One of the most recognized insects is the praying mantis. How did the insect get such a pious name?

The word "mantis" is based on the Greek word meaning prophet or soothsayer. While in wait for a victim, the mantis takes a rather pious stance which no doubt suggested its name—praying mantis. The European mantis even has a scientific name descriptive of the posture, *Mantis religosa*.

The praying mantis is no pacifist. It would probably be more appropriately called the "preying mantis" for it is an effective predator. Most mantids are colored in such a way as to camouflage their presence. They hunt by waiting. They have small heads and are among the few insects with a distinctive neck. The neck allows them to turn their heads so they can better spot their prey, Their large compound eyes give them great vision for detecting movement and judging distance.

All mantids are carnivorous and feed on almost any insect including flies, crickets, moths and even wasps. Mantids are well-equipped to catch their food. Their forelegs are modified into a fearsome insect-catching device that functions like a pocket knife. These raptorial forelegs include spines to aid in holding a victim caught in ambush.

Once mantids have procured a meal, they eat with the aplomb of a sophisticated diner at the finest of restaurants. However, their dining manners belie the appetite of a woodchopper. They can consume as many as twenty flies in one day.

It is this appetite and food habit that makes the praying mantis a beneficial insect. The insect is a friend of gardeners everywhere. However, mantis populations are seldom high enough to destroy large populations of pest insects. Therefore, many gardeners give mother nature a little hand by purchasing and placing mantis egg cases in gardens and flower beds.

The purchase of praying mantis egg cases may or may not rid a garden of undesirable insects. However, the act of doing so makes most gardeners feel good. Besides, they take great pleasure in seeing a mantis in the garden. Especially if that mantis happens to be chowing down on some other insect!

Sticks That Walk

What's the difference between a walking stick and a stick that walks? Six legs!

When we think of a walking stick, we usually think of a staff or cane that aids a person in walking. However, there are walking sticks of the insect variety that walk with six legs.

Walking sticks that are commonly found in the United States resemble twigs. They are wingless, slow-moving insects that feed on trees and shrubs. Their version of a twig look-alike contest is a matter of survival. They blend into their enviornment so that potential predators can't see them easily. Small birds perched upon a walking stick are sitting on a potential meal but often don't know the difference.

Insect walking sticks vary in size. The largest insect found in the United States is a walking stick that reaches a length of 6 to 7 inches. This insect is found in the South and Southwest. Most walking sticks are tropical and those that live in such climates can attain an adult length of about a foot, which is appropriate considering their name.

In some instances, walking sticks have covered trees and consumed all the foliage. Because of their protective coloration most people will not see them. To the untrained eye, they are just another bare twig.

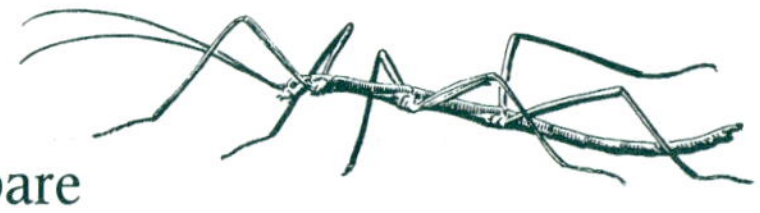

Walking sticks have a single generation per year. Mature female walking sticks drop their eggs wherever they feed, leaving them to endure the winter season on their own. After winter, the little twig imitators hatch in a paradise of fresh, green leaves which they will eagerly devour.

The next time you focus your gaze upon what you think is a twig, and it walks away, don't fret. You've just had an insect encounter of the oddest kind. Walking sticks are one of the many bizarre insects that inhabit the earth.

Hoppers, Hoppers Everywhere

Ah, July! That time of the year when ol' Sol the sun, almost without human notice, begins a seasonal retreat toward the southern horizon, and the days begin to shorten ever so slightly.

To most folks, July signals the onset of, as Nat King Cole immortalized in song, "Those lazy, hazy, crazy days of summer." However, the declining minutes of daylight stir within plants and many animals a sense of biological urgency. Plants flower; some animals lay in a winter's food supply; others just add on a layer of fat; and many insects rush to lay eggs.

No insect is more closely associated with the dog days of summer than grasshoppers. If baseball players can lay claim to the title of "Boys of Summer," then surely grasshoppers are the "Bugs of Summer!"

Grasshopper egg masses, oviposited the previous fall in the soil, hatch in June and July. As summer days get drowsier, the grasshopper nymphs—typical teenagers that they are—are eating up a storm.

By August, mature grasshoppers begin the process of mate selection. Many grasshoppers attract mates by producing a sound, appropriately named "calling." This sound, called a "song," is most often produced by males in one of two ways. The short-horned grasshoppers—those with short antennae—make sounds by rubbing their hind leg across their wing. The long-horned grasshoppers rub one wing against the other.

It is a bit misleading to term these sounds "songs" though, because the sound has no pitch. It sounds a bit like rubbing two pieces of sandpaper together. For most of us, this sound is relegated to the status of background noise, and it is easy to forget or ignore.

Grasshopper songs are one of the most prominent features of late-summer days. At almost any time of the day, three or four species of grasshoppers can be heard. They, like birds, can be identified to species by their calls.

It is probably this singing that prompted Aesop to pen his fable of the grasshopper and the ant. To Aesop, it must have seemed that singing away the last days of summer was a foolish thing for a grasshopper to do, especially compared to the ants, who work industriously to store the winter's food supply. But the singing of the grasshopper should probably stand to remind us that its work is done. The grasshoppers' late-summer song is really a last laugh. For in the soil are grasshopper egg masses, just settling in for a long winter's nap in preparation for next season's feast.

Other Insect Orders

1. What kind of an insect was archy?
2. Describe the process of how dragonflies catch food.
3. In what type of habitat would dragonfly immatures be found? What is their food?
4. Where would winter stoneflies be found during the winter months?
5. Under what weather conditions do grasshoppers cause the greatest problems?
6. Why is it appropriate that mayflies are classified in the insect order Ephemeroptera?
7. What is the ecological role of mayflies?

UNIT 4

Insect Biology

CHAPTER 13

Insect Reproduction

Vive La Différence

In the insect world, male and female insects are different in ways other than the obvious physical attributes.

In the insect version of the dating game, insect sounds for attracting a mate are always produced by the male. The songs of crickets, grasshoppers and katydids emanate from an all-male choir. And the not-so-melodious shrill of cicadas results from the vibrations of a drumlike membrane on the abdomen of the males.

In fact, ancient Greeks recognized that male cicadas sang, while females were mute. The Greek philosopher Zenarchus summarized this fact in his comment, "Happy are the cicadas' lives for they have voiceless wives."

Female insects have their own ways of dealing with the opposite sex. Many females produce mating perfumes called pheromones. Upon release, these chemicals, which are emitted from specialized organs, attract males from considerable distances. Some of the large moths may travel up to 50 miles while following the pheromone trail of a female. Although females dominate the insect perfumery business, a few male insects, such as some of the bark beetles, produce pheromones, too.

Following mating, many male insects go their merry way while the task of laying eggs and caring for the young is left to the females. But some female insects don't let the males off that easy.

For example, the giant water bug female captures a male and attaches her eggs to his back for him to carry until they hatch. Other females—such as the praying mantids, some dance flies and even some fireflies—make a meal of their mate.

In some insect species, the number of males are less than 50%. The engraver beetle populations are made up of a ratio of 200 females to each male. And some aphids have eliminated males entirely. These females reproduce through a process called parthenogenesis, without mating.

So in the insect world, it may be the males that do the singing, but more often than not, it is the female that has the last laugh!

Insects Don't Add; They Multiply

Ever wonder how there can be so many insects so quickly after the spring thaw? It could be magic, or spontaneous generation. But it isn't either.

It has to do with the birds and the bees—mainly the bees! Bees and other insects are very good at laying eggs in very large numbers. Scientists believe that rapid reproduction is one reason for the phenomenal success of insects on the earth.

Consider the honey bee queen. In the warm months of June and July, she can lay between 1,500 and 2,000 eggs each day. Since she must deposit each in a cell of the comb, that production represents a good day's work. You might say she's "busy as a bee" during that time. A honey bee queen will produce around 100,000 eggs during her average 3-year life span. Not a bad record.

But honey bee egg production pales in comparison to that of the large queen termites of Africa. Such queens, real egg-laying machines, may lay as many as 36,000 eggs in 24 hours. That's about 25 eggs a minute and totals over 13,000,000 per year. Such queen termites are believed to live between 50 and 100 years. That figures out to be over a billion eggs produced by one queen.

While the old house fly is no queen termite when it comes to laying eggs, she isn't a slouch either. Average egg production during the life of a house fly is probably around 800 eggs. The record number of eggs measured for one fly is 2,387. But the success of the house fly is not solely due to the rate of egg-laying; it is also due to the speed with which the insect goes through its life cycle.

The same is true for the fruit fly, an insect that almost everyone recognizes. Fruit flies seem to magically appear around overripe vegetables or fruit. The appearance of large numbers of fruit flies in a short time shows how fast this insect develops. Under ideal conditions, fruit flies may produce 25 generations per year.

Of course, all insects that hatch don't survive to reproduce. Many insects die of starvation or exposure to lethal temperatures or are consumed as food by other animals. Among insects, many are hatched so that a few survive. You could say that in the insect world, the battle cry is, "The more, the merrier!"

FIGURE 13.1. *Insect Eggs.*

Insect Eggs

Insect lives are frequently described in terms of a cycle: egg to larva to pupa to adult and back to egg. So the age-old paradox frequently asked of our fine feathered friend, the chicken, might also be relevant to insects. Which came first: the adult or the egg?

Every insect begins life as an egg. Most insect eggs are laid by the adult before the embryo is fully developed. Therefore, the egg doesn't hatch immediately.

Hatching time is usually short, normally within a few days. Since insects don't incubate their eggs, hatching time is determined by the environmental temperature. For instance, the eggs of the Oriental fruit moth will hatch in three to six days in warm weather, but may take up to 40 days in cooler weather.

Some insects hibernate in the egg stage. In this case, the egg may not hatch for 6 to 10 months. Grasshoppers and corn rootworms are such insects. And eggs, such as of the walking stick, may remain unhatched for two years.

Fertilization is generally necessary to produce an insect egg that will hatch. Some insect females have, however, eliminated the middleman in the reproductive process. Insects that lay unfertilized eggs include some bees, ants, social wasps and many aphids. In the social insects, unfertilized eggs develop into males while fertilized eggs become females.

Insect eggs come in various shapes. Some are flat and scale-like. Eggs of the European corn borer look like fish scales, half covering each other on the leaf of the corn. Others look like miniature wine barrels placed neatly on a dock awaiting export. Some of the stinkbugs follow this pattern of egg placement.

The shell of the insect egg can be perfectly smooth. However, some insect eggs are sculptured in striking markings that are the envy of human artists. Generally, the thickness of the shell of the insect egg gives a hint of the environmental conditions it must face. Eggs that hatch shortly after they are laid have thin shells. Eggs that must endure the winter tend to have thicker shells. An exception is when the egg is laid in some protected place, such as in the soil or under the bark of a tree.

Some insects cover the egg mass with special materials. Many moths cover the egg mass with hair from their bodies. The female bagworm deposits its eggs in the bag which served as her home. The praying mantis secretes a brown, frothlike substance in which to enclose her eggs and attaches the mass to a twig or wire fence, where it remains for the winter.

While it is not possible to answer the question of which came first, it is true that the insect egg must survive if there are to be adult insects. Diversity in the shape, size and placement of eggs suggests that insects have opted not to put all their eggs in one basket!

Insect Songsters

Insects rival birds as the animal world's most noteworthy singers. The rasps, shrills, clicks and chirps of insects aren't as melodious as the tones pro-

Say It with Perfume

It's not easy to find someone with whom to share your life. Especially if you are an insect. In fact, most adult insects devote their lives to that most ancient of biological imperatives—finding a mate. Insects don't have computer dating lists, lonely-hearts clubs, singles bars or even a six-legged version of a matchmaker to help in the mating game. Each insect is on its own when it comes to this business of love and marriage.

Insects employ a variety of activities to attract potential mates. Some, like crickets and katydids, are crooners. Others like the fireflies are specialists in aerial fireworks. Still others "say it with perfume." Yes, for many insects a good scent is worth a thousand sweet words.

Insects that use perfume to announce their availability include butterflies and moths. Technically, these velvet-winged suitors use chemicals called pheromones, which are produced in the insect's body. When released into the environment, these odors are irresistible to the opposite sex. Most pheromones are produced by female insects, but in some species the male is the one in the perfumery business.

How does this all work? The female produces the pheromone and releases it into the environment. To make sure that the chemical gets carried toward potential suitors she flaps her wings about. This creates a plume of air that wafts downwind carrying the sweet scent with it. This process is appropriately named "calling"—sort of the insect equivalent of "Hey, good lookin'"! Males who happen to be downwind at the time will detect the pheromone—that's why male moths have fuzzy antennae—and follow the trail upwind to the source. At that point, one is to assume he politely introduces himself and says something like, "Nice pair of legs . . . pair of legs . . . pair of legs."

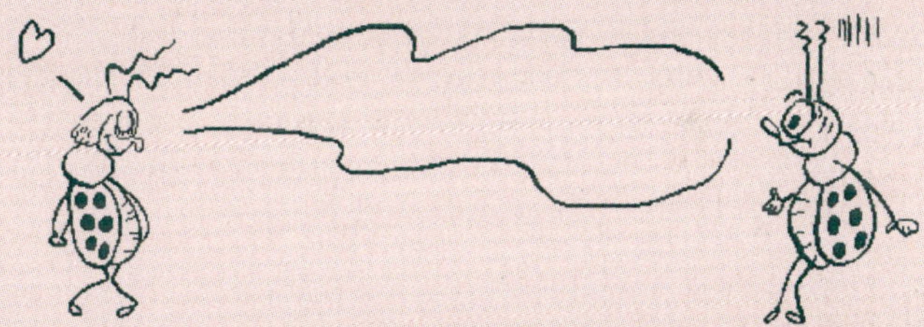

Pheromones are complex chemicals. For instance, the pheromone produced by the European corn borer is technically termed (Z)-9-tetradecen-1-ol formate. But to a corn borer, that chemical probably rivals the world's sweetest perfume.

We humans once had pheromones of our own. Then we became civilized and began taking baths, and the natural pheromones went down the drain. Thinking individuals that we are, we recognized the importance of such odors and adopted a technological solution to the problem. We make our own pheromones and splash them on. We call them perfumes, colognes, after-shaves and such. To add to the mystique, we give our pheromones fancy, and sometimes seductive, names. And pay high prices to buy them.

Do you suppose that we humans should admit that we adopted an approach to selecting our mates that was originally perfected by insects? Why not? It certainly makes good scents!

duced by birds, but, by sheer volume, the insects frequently dominate their avian competition in nature's chorus.

Unlike birds, insects use a variety of mechanisms to harmonize. Some, like mosquitoes and bees, produce sound by the vibration of their wings.

Such sound is incidental to the activity of these insects. However, the mosquito's whine has come to be a warning to its human victims. Likewise, the buzzing of bees is widely recognized as a danger sign in the animal world. Some flies even mimic the sound for their own protection. And the sound is so familiar that the composer Rimsky-Korsakov used violins to represent it in his popular "Flight of the Bumblebee."

The majority of insect choristers produce sound by stridulation. That is, they rub one body part against another. Some grasshoppers produce sound by rubbing the edges of their wings together. Others "fiddle" by rubbing the back leg across the edge of the wing.

But all insects aren't fiddlers. Some are drummers. Cicadas are the percussionists of the insect world. The drum of the cicada is a membrane located on the abdomen, which is vibrated by the use of a complicated set of muscles to produce the well-known hum of the cicada.

Insect sounds vary throughout the season and over the course of the day. For example, many grasshoppers sing during the day but become silent

Singing and Stinging Insects Can't Hide Their Sex

The general rule in the animal world is that females and males of the same species look different. Sometimes the difference is in size or color. Birds are often examples of color differences between sexes. Differences also might include the presence or absence of horns. This is the case in some deer and sheep.

Scientifically, a different look between males and females of the same species is known as sexual dimorphism. And the differences are—you guessed it—related to reproduction.

To the untrained observer, male and female insects look the same. In some instances, even experts on insects can't tell bug girls from bug boys. Such a determination would involve dissection of the insect to look at the internal workings.

There are obvious differences between the sexes in some insects. Sometimes the difference is most noticeable in behavior. For instance, any adult insect that stings is a female. That's because the stinger is the insect egg-laying device known as an ovipositor.

Bees and wasps are well-known insects with stingers. However, the females don't fly around with their stingers out in the open. So, for most people, getting stung by a bee or wasp is the only sure way to know that the insect is a female. To make matters worse, some male wasps, like the cicada killers, actually act like they are trying to sting. This female impersonator actually goes so far as poking you with the end of his abdomen!

The ovipositors of female grasshoppers, katydids and crickets are obvious extensions of the abdomen. These ovipositors resemble curved knives, which is how they function, since they are used to cut plant tissue or soil to deposit eggs.

In some insects, males and females look entirely different. One of the most obvious differences has to do with wings. Males have wings and females do not. This is true of some fireflies. There are also some moths where the male is the only sex with wings. This is true of that pest of evergreen trees and shrubs, the bagworm, where the female stays in the bag. The infamous gypsy moth female is also wingless.

In some of the large walking sticks that thrive in tropical regions, the males are winged while the females are wingless. In these insects, the females are also much larger, making the males the "Jack Sprats" of the insect world.

Size-wise, praying mantis males are also lightweights when compared to their females. So when a praying mantis female decides to devour her mate, the poor male is fighting up a weight class or two.

Many male moths have fuzzy antennae compared to females of the same species. That difference is related to detecting the chemical mating perfumes released by the female. The chemicals are trapped in the fuzzy antennae of the male and tell him somewhere upwind is a female looking for a mate.

Male mosquitoes also have fuzzy antennae. In this case, the antennae allow them to pick up the sounds of the wingbeat made by female mosquitoes. That is the same sound that we hate to hear in our ears. To the male mosquito, it is music to his antennae.

Some of the most obvious sexual differences in the insect world are associated with stag beetles. The males of these beetles have large horn-like projections that are used in battles between the males. Like their mammal namesakes, the males butt and push each other in contests for territory and females.

The male monarch butterfly has a black spot on each hind wing, something that the female lacks. The black spot is a scent gland that produces a chemical attractive to the female of the species.

Male insects are the songsters of the insect world. Sounds of katydids, crickets, grasshoppers and cicadas are produced by the male to attract their females.

It is not always easy to tell a female insect from a male. But of this we can be sure: if it stings, it is female; if it sings, it is male!

Most Insects Exhibit Poor Parenting Skills

In the animal world, parenting is something that comes naturally. It's in the genes. Mammal mothers, like deer, dogs, sheep and cows, lick the newborns and nuzzle them into positions to nurse. We call it "maternal instinct."

Adult birds brood the eggs and carry food to the hatchlings in the nest. At night, the mother provides the young with warmth and protection.

In the case of predators like foxes, the parents even teach the young to hunt. The same is true of birds. Outside the nest, the adults show fledgling birds the finer points of rustling up a meal. That is what the clucking and scratching of a mother hen with a brood of chicks is all about. The hen is teaching the chicks how to find food.

While parenting comes naturally to most creatures, the two most successful animals on Earth aren't good at it. Humans and insects rank up there among the worst parents in the animal world.

Of course, calling humans and insects poor parents is a value judgment. After all, both groups are highly successful animals. But humans seem to have to learn to parent. Otherwise, why would we buy so many books on parenting? Since Dr. Spock pioneered such books, literally dozens of titles on the subject have appeared on booksellers' shelves around the world.

Insect parenting varies from minimal input to a sophistication that even we highly evolved humans try to emulate. Most insects leave their offspring to fend for themselves, but even these try to give their kids a good start in life.

The simplest approach to the insect version of child rearing is to do nothing. Some insects simply drop eggs as they feed. Such is the case with some walking sticks that drop their eggs from the trees. Some cockroaches deposit their egg cases, called ootheca, as they run along. Mayflies actually lay eggs as they die, wherever that may be.

A common insect approach to bringing up junior is to place the eggs in a desirable location. That generally means in a place where the newly hatched insect will find food. For instance, many butterflies and moths will fly around looking for a suitable food plant for their offspring. When the female finds such a plant, she will lay an egg. When the egg hatches, the caterpillar doesn't have to move far to find food.

Sometimes the egg is just placed among food items. Ladybug mothers-to-be deposit their eggs close to aphids. Since ladybugs and their immatures feed on aphids, the baby ladybug is normally surrounded by potential meals.

Many flies feed on dead or decaying plant and animal matter, so the eggs are laid on—you guessed it—dead stuff. That explains why you see flies buzzing around road kill; many are laying eggs. The resulting maggots begin feeding on the rotten flesh.

Placing the eggs in a desirable location by insects would be equivalent to humans leaving their children in a fast-food restaurant. The food would be there; the kids would have to find it.

Some insects provide the kid's food in bulk—enough food to meet the lifetime needs of the immature. The mud dauber wasps that provision the nest with spiders or caterpillars use this approach. Enough of the prey items are placed in each cell to feed the larvae. Cicada-killer wasps provide a single cicada for each wasp larva. Leaf cutter bees and carpenter bees store enough pollen in a nursery cell to meet the needs of each baby. This would be like a human parent giving a lifetime certificate for the fast-food restaurant.

Some insects, like the paper wasps, supply food as the offspring need it. They are called "progressive provisioning insects" and could be compared to humans who provided coupons for food as needed.

Some insects actually turn over the whole process of feeding the offspring to workers specialized in child care. This is the way of life for social insects, like ants, termites, bees and wasps. In this case, the workers prepare the food and feed the kids. Sounds very much like a family that hires a nanny to care for the kids, doesn't it?

Battle of the Sexes—Insect Style

With humans, it began with Adam and Eve. Insects, it seems, have been at it even longer. Yes, the battle of the sexes is not just for people!

Some of the really big gender behavioral issues that bug humans don't bother insects. For instance, whether the toilet seat should be left up or down isn't an insect thing. Neither is the propensity to assume that males won't ask directions. Shopping and moving furniture? Not an issue with insects.

But there are behavioral differences in insects that are sex-related. For instance, mate selection. In the insect world, a common approach to mate location is that the female allures and the male pursues.

For many insects that means using a pheromone, a perfume, to attract a mate. In insects, the use of pheromones is a decidedly distaff thing. With a few exceptions, pheromones are produced by female insects.

But then, that hardly sets insects apart from humans. If you aren't convinced, just take a look at the cosmetic section in your favorite store. You will find that the selection of artificial odors for women far outnumber those for men.

Sounds are also used by insects to attract mates; mostly, it is the male insect that produces sounds for the purpose of attracting a female. The process is known as singing. The exception to males producing the mating sound is in some flies and mosquitoes. In these insects, the hum produced by the wings of the female during flight attracts the male of the species.

Many people assume that singing katydids are female. Perhaps, it is because the common name of the insect, katy, suggests it. Some poets have also been confused about the sex of the singing katydids. Oliver Wendell Holmes in his poem "Katydid" states:

"Thou art a female Katydid
I know it by the trill
That quivers through thy piercing notes,
So petulant and shrill."

Philip Freneau also assumes that this insect is female. In his poem "To a Caty-did," he uses female pronouns in reference to the insect: "Hear her singing in the shade" and "Caty now will do her best, All she can, to make you blest." But poets aside, any katydid, grasshopper or cricket that sings is a male!

Child care, where it occurs in insects, is mostly left up to the females. The eggs may just be hidden in a safe location, such as the soil or inside the stem of a plant. Some insects, such as the cockroaches and praying mantids, place the eggs in an egg case. But whether protected or not, when the young insects hatch they are on their own.

In social insects, the workers, which are females, do all of the kid care. They feed, clean up after and pamper the larvae until they turn into adults.

In fact, all of the work of the social insect colonies, including honey bees, bumble bees, ants, termites and some wasps, is done by female insects. Even the head of the colony is a female. Called the queen, she lays the eggs and generally is in charge of the rest of the colony, most of which are females. Each colony does have some males. These males, in the bees, are called drones.

Drones have only one function, to mate with a queen. In between flights to search for queens, the drones lounge around the hive and are fed by the workers. A good life, except there is a catch. When winter rolls around, the females of the hive perceive drones as a liability and toss them from the hive.

Now being tossed from your home is the ultimate in the feud between the sexes. It kind of makes being browbeaten for not stopping to ask directions seem mild. Maybe it has to do with the lack of a pheromone!

at nightfall. Many crickets begin to warble when the sun goes down.

Insects sing primarily for the purpose of communicating with others of their species. For instance, some insects court a potential mate by crooning a sweet song, a practice the insects share with birds and, in some instances, humans.

Many people over the years have enjoyed the harmonic renditions of insects. Some Oriental cultures keep crickets in cages just for the beauty of their singing. Indeed, a cricket chirping in the house is considered good luck. Such an idea was noted by Charles Dickens in his poem "Cricket on the Hearth."

However, not all people regard the sound of insects as a thing of joy. For instance, an innovative entomology student once proclaimed on a test that cicada killers kill cicadas because "they can't stand the noise!"

Insect Reproduction

1. What is unusual about the egg laying habits of the giant water bug?
2. In the social insects, unfertilized eggs hatch. What is the sex of the individuals from these eggs?
3. In general how does the covering of insect eggs reflect the environmental conditions the egg will encounter prior to hatching?
4. What are insect "perfumes" called?
5. What is the name of the process in which mate-attracting chemicals are released into the air by an insect?

CHAPTER 14

Food Habits of Insects

Food is essential to the growth and development of any organism. Insects use a wide range of food sources. Insects can be considered to be saprophagous, phytophagous or zoophagous. Saprophagous insects feed on dead organic matter. Phytophagous insects feed on living plants and zoophagous insects feed on living animals.

Saprophagous insects could also be called scavengers. Some of the insects that are saprophagous are cockroaches and house flies. Both are general feeders on either plant or animal matter. Collembolans (sometimes called springtails) feed on soil humus. Termites, however, are restricted to feeding on dead plant tissue (occasionally in semitropical regions they may attack living plants). Carrion feeders such as flesh flies and burying beetles are restricted to animal food. Very specialized feeders in this category are the insects that feed on animal dung—the Coleoptera and Diptera are common in this category. One of the most well known is the dung beetle, sometimes called tumblebug, that rolls balls of cow manure to bury and feed to its young.

Phytophagous feeders include leaf feeders like grasshoppers. Also in this category are root feeders like grubs, leaf miners, and stem borers (like the European corn borer). The gall makers, the fungus feeders and the sap suckers such as the plant bugs and aphids are common members of plant feeding insects.

Zoophagous insects include parasites of all kinds. Fleas and lice feed on higher animals. Ichneumon wasps and flies are parasites on other insects. Predators—they consume more than one animal during their life—include ladybird beetles, some wasps and some bugs. Included in this category are the blood feeders like mosquitoes and deer and horse flies.

It Takes Gall

Almost everyone has, at one time or another, noticed a deformity in plant growth called a gall. These often grotesque plant abnormalities can be caused by insects. Gall-causing insects lay their eggs in plant tissue, and the developing larvae cause the plant to respond with abnormal growth, known as a gall. Some beetles, moths, flies and wasps are among the insect gall-makers.

Galls have been recognized by humankind for centuries, and in fact, galls are better known than the insects that produce them. The galls are numerous and showy; their insect residents, on the other hand, are quite small and difficult to see.

From times of old and even today, galls have been used to produce tannic acid and dyes, including one dye called Turkey Red. The best permanent inks have for years been produced from galls. The Aleppo gall, found in Eastern Europe and Western Asia, is used for this purpose. In some places, the law requires that permanent records be made with ink derived from galls. Such an ink is used by the United States Treasury and the Bank of England.

The word "gall" has a double meaning in the English language. The Latin word "galla" meant a gall-nut, the plant growth abnormality. The Anglo-Saxon word "gall" meant bitter, hence the word "gallbladder" to identify the bile-holding sac associated with the liver. Shakespeare, no doubt, had both meanings in mind in "Twelfth Night" when he had Sir Toby Belch tell Sir Andrew to use "gall enough" in the ink to pen a challenge to duel Viola.

Gall insects attack over one-half of all plant families, and almost all parts of the plant are subject to infestation. In general, gall-makers produce the same type of gall on different plants.

Galls are generally named according to the way they look. For instance, we have the hedgehog gall, which must look a bit like a hedgehog. Many galls remind folks of fruits and vegetables, including the apple, potato and pea galls of oak. We also find spindle, hairy and spiny galls on oak. And there are button galls, cup galls and red sea urchin galls. Bullet galls and vase galls have been known to occur on the same plant as the wool-sower galls.

One of the most interesting of the galls is the jumping oak gall. When this gall falls from its oak tree host, the larvae inside cause the gall to jump. The popping activity of the jumping gall is similar to that of Mexican jumping beans, which is also caused by an insect inhabitant.

Even in the insect world, a little gall can sometimes attract attention!

Food Habits of Insects

1. What is a purpose of insect sounds other than for attracting a mate?
2. What is the food of saprophagous insects? Name three such insects.
3. Grasshoppers and European corn borers exhibit what general food habit?
4. Name three types of insects that are zoophagous.

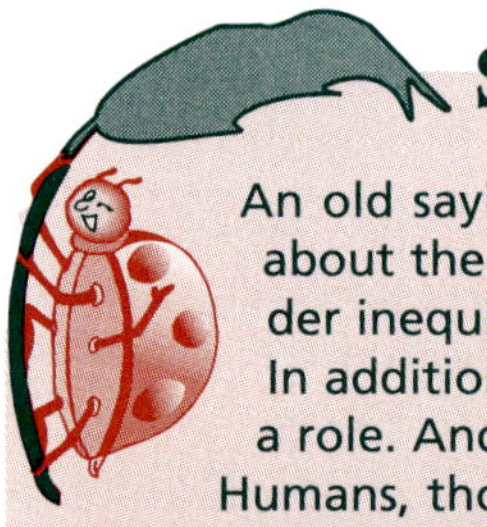

Some Insect Pests Are Nature's Helpers

An old saying holds that "What is good for the goose is good for the gander!" I don't know much about the wants and needs of male and female geese. But without prompting a firefight about gender inequities, I suspect that a goose and a gander don't always have the same needs.

In addition, the idea of good and bad is meaningful only in human terms. In nature, everything plays a role. And that role is neither good or bad. So it is with the insects of the world.

Humans, though, have other ideas when it comes to classifying things in nature. We like to look at nature in terms of "What did you do for me lately?" So, insects are either good or bad depending on how they affect what we want.

An insect that feeds on plants, in ecological terms, would be called an herbivore. Herbivore does not connote goodness or badness. It is a non-judgmental name for what the creature does. In this case, eat plants.

We humans claim some plants as our own—plants like crops, lawn grasses, flowers and trees for wood or decoration. When we claim a plant and an herbivore shows up to have a meal, we get bent out of shape. We even have a name for such a creature—pest! We then do what we can to make life miserable for the pest.

Fickle creatures that we are, we sometimes find insects to be both good and bad, for instance those insects known as dermestid beetles. These insects belong to the family Dermestidae and play an important role in nature as recyclers of the skin and hair of dead animals. That is the basis for the name dermestid. It is from the Latin root derm, which means skin.

Dermestids are also called hide beetles or carpet beetles. The latter name suggests the source of the problem with these insects. As nature's cleaner-uppers, they take their job seriously. Material from dead animals, like wool in mittens, socks, suits, or carpets, leather or fur in coats, or animal-skin rugs is a signal for egg laying. When hatched, the larvae begin to feed on the natural material and that is when we end up with holes in our best woolen clothing. Of course, the owner of the clothing doesn't appreciate that the insect is just doing what nature ordained it to do!

Resourceful humans have found a way to use the dermestid beetles. These insects are also called museum beetles. They have that name for good reason. Dermestids are used to clean up bones that are destined for museum displays. No one wants to look at a display skeleton that has bits of dried flesh and skin attached. So museum curators clean up skeletons for display by placing the bones in containers that have a population of some dermestid beetles. The beetles do their work. The bones are nice and clean, and ready for display and viewing by museum visitors.

But in a museum setting, the dermestids pose a severe hazard. After all, many museum exhibits feature stuffed animals that have their natural skin and fur or feathers. So dermestid beetles can destroy the exhibit by feeding.

Dermestid beetle feeding can also damage insects in collections. The dead insects are as good a food resource as are skin and hair for immature museum beetles. Many a private insect collection has been turned to dust and a few odd wings by the feeding of these beetles. That is why insect collectors frequently treat their insect specimen drawers with insect repellents. The chemicals are used to make sure that the only insects present are dead ones!

It just goes to prove that humans are fickle animals. We can't make up our mind about the dermestid beetles. We want them to get rid of the road kill. We don't want them to get rid of our woolen socks. We want them to eat the skin from museum skeletons. We don't want them to eat the dead insects in museum drawers. At least, dermestid beetles don't worry about split personalities!

CHAPTER 15

Surviving Winter

Where Have All the Insects Gone?

Each year as fall marches inevitably toward winter, nature prepares. Temperatures begin to drop and so do the leaves from trees. Waterfowl wing overhead in their annual southern migration from breeding grounds in the north. Swallows, bluebirds, and red-wing blackbirds depart to make way for the birds of winter.

And the insects disappear. Yes, the teeming hordes of insects that occupied every nook and cranny just a few weeks ago are seen no more. The cricket, the cicada and even the mosquito have fallen silent.

To borrow, and slightly modify, a line from an old Peter, Paul and Mary song, "Where have all the insects gone, long time passing?" If the answer is "to graveyards every one," it is a bit misleading. To be sure, many of the insects of last summer have died. But their offspring shall return!

Insects have solved the problem of winter in several ways. Some hibernate. Ladybird beetles, Mexican bean beetles, bald-faced hornets, yellow jackets and cluster flies are among the many that hibernate as adults. These insects seek sheltered places to gain some protection from the winter. They can be found under rocks, in leaf litter, even under the eaves and in the attics of our homes.

Other insects spend the winter as larvae. The larvae of June beetles, known as white grubs, just dig deeper in the soil as the temperature drops. Many caterpillars, such as the wooly bear, crawl into secluded places for their long winter's naps. Most, however, make some provision to keep from freezing. Some reduce the water content of their bodies and replace it with glycerol, the antifreeze of the insect world.

The European corn borer spends the winter as a larva in corn stalks. A fact of interest to ice fishermen who frequently dig the borers from their hiding places to use as bait.

Some insects avoid the winter by remaining in the egg stage. The mosquito, corn rootworm and praying mantis leave their eggs out in the cold.

Other insects solve the problem of Midwest winters by leaving. Yes, like some of the human species, they go south for the winter. Many of our insect pests repopulate Midwest areas each spring from breeding grounds in the south. Many aphids, cutworms and armyworms use this strategy for dealing with winter. Another insect that migrates back each year is the monarch butterfly.

Insects have used a variety of tactics to solve the problem of Indiana winters. While insects may be gone, they should not be forgotten. For next spring, when the temperatures begin to warm, the fire of life will be rekindled in tiny insect hearts everywhere. Like the swallows to Capistrano, they shall return!

Southward Bound

The autumnal equinox is the official beginning of fall. 'Tis the season when pumpkins sometimes wear a mantle of frost in early morning. When the fading green of the leaves of summer expose bright but fleeting hues of red and gold.

Fall is a time of harvest and quiet beauty. Mother Nature removes her summer foot from the accelerator and life begins to coast. Slowly at first, then more rapidly, the earth's life cascades toward the barrenness of winter.

Fall is a harbinger of death to many of the insect world, but this is not so for the monarch butterfly. One of the best known of North American insects,

Insect Antifreeze

November is the time to winterize homes, cars, plants and even dog houses. Surviving winter demands preparation. It means caulking, putting up storm windows and doors, covering winter-sensitive plants with mulch, insulating dog houses and possibly getting a flu shot.

What about insects? What do they do when Old Man Winter blows his icy breath across the landscape? Well, they also winterize.

Winterizing for an insect is much like the process we go through to winterize a car. We add antifreeze to the car. Insects add antifreeze to themselves. If the liquid in the cooling system of a motor is allowed to freeze, the expansion during the process will break the radiator and hoses. The same is true of the liquid in insects. If it is allowed to freeze, the crystals that form will destroy the cells and tissues of the insect and cause death.

By adding antifreeze to an automobile's cooling system, the freezing point of the liquid is reduced so that freezing doesn't occur. This protects the system from damage.

As winter approaches, some insects use a similar antifreeze approach. In preparation for freezing temperatures, the water content in the insect is reduced. The water is replaced with glycerol, a compound similar to the glycol that's used in antifreeze. Therefore, like the car, the insect is winterized. An insect's biological system is shut down and is protected from freezing during the cold winter months.

When the days become longer and the temperatures creep upward, the insect reverses the process. Glycerol is broken down and replaced with water. The insect is ready to resume normal activities.

In her poem entitled, "About Caterpillars," Aileen Fisher begins with this stanza:

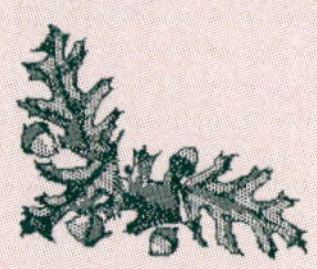

"What about caterpillars?
Where do they crawl
when the stars say, 'Frost,'
and the trees say, 'Fall.'"

Good question. But the important thing about an insect surviving winter is not where they crawl, but what they do when they get there. The successful insects winterize with antifreeze.

it avoids winter by migration. As fall approaches, the adult butterflies begin a truly miraculous journey to their overwintering sites.

Mountainous areas in Mexico and California provide the ideal environment for these travelers to survive the winter. Presently, only two major sites have been identified by entomologists as meeting the monarch's requirements. In fact, because deforestation endangered one of these butterfly spas, Mexico declared it a protected area.

As these insects travel south, they create major excitement when they stop to spend the night and congregate on a single tree. These trees, sometimes called butterfly trees, are the insect equivalent of a Holiday Inn. Once at the overwintering site, the marvelous monarch feeds on nectar from flowers to build up energy reserves for a return flight.

As spring approaches, the overwintered monarchs begin to wing northward, as their ancestors have done for untold centuries before. During the spring journey, they mate and lay eggs on milkweed plants.

The sight of monarch butterflies fluttering lazily in a southerly direction across an October meadow gives one pause for thought. That rather fragile insect with thin membranous wings covered with soft scales is in the middle of a journey that can cover up to 2,000 miles before it ends. It is on a journey to a location where it has never been, but to which it is unerringly guided by some genetic code.

It is truly a miracle of the biological world.

Plant Chompers

Gardeners know it. So do farmers. It's a little song and dance number that insects do—"the plant chomp."

Eating plants is popular in the insect world. Almost all butterflies and moths eat plants during their caterpillar days. Most crickets, grasshoppers and true bugs are plant feeders as well as many beetles, some flies and a few bees.

This chomping of leaves, chewing of stems and sucking of sap is not good for the plant. So plants have developed defense mechanisms to combat the insect menace. Some plants have physical barriers such as spines and hooks to discourage insects from chomping them. Others contain noxious chemicals that are harmful to insects and animals in search of a plant meal.

These noxious plant chemicals, sometimes called phytotoxins, are effective inhibitors of insect feeding. Some phytotoxins have been discovered to have uses other than keeping insects from feeding.

Long before mankind had available synthetic insecticides, it was discovered that pyrethrin, an extract from chrysanthemums, killed insects. Other such plant-derived insect killers include rotenone and nicotine.

Some of these chemicals have medicinal properties. Quinine from cinchona bark was used to treat malaria. The milkweed plant contains a chemical called cardiac glycoside, because it is useful in treating certain kinds of heart diseases.

Other phytotoxins include mescaline from the peyote cactus and morphine from the poppy. The active compound in marijuana is cannabidiol; the coca plant produces cocaine. Both compounds function to protect the plants from insect attacks. The nice aroma of our morning cup of coffee portends the stimulatory effect of the caffeine. While in the coffee plant, caffeine is a deterrent to insect feeding.

It is interesting that many chemicals that protect plants from insect feeding work as stimulants or hallucinogens when consumed by humans. Could it be that insects that ingest such chemicals are too nervous to eat? Or perhaps insects just don't care enough to worry about food? Or it may be that even a lowly insect is too smart to eat what is bad for it!

On the Winds of Spring

Ah, the southerly breezes of spring! Those warm and sometimes turbulent air masses signal that Old Man Winter is losing his grip for another year.

Spring breezes seem to spontaneously generate children with kites, and enthusiastic and overly optimistic gardeners, April showers, and hordes of insects. But the winds of spring also bring the promise of teeming multitudes of insects. Flying, jumping, crawling, sucking, munching insects.

To some insects, the onset of spring is merely a signal to emerge from their winter hiding places and begin the frenzy of insect activity associated with summer. To other insects, however, the spring weather fronts actually provide transportation to summer feeding grounds.

Many insects cannot overwinter in northern areas, but they survive very well in the more moderate temperatures of the Gulf Coast states. As the spring weather fronts begin to generate, they pick up more than moisture from the Gulf. The fronts also pick up insects that happen to be flying around at the time as well as those insects that are waiting to catch a ride northward.

Many spring weather systems are loaded with insects. In fact, so common are insects in weather fronts that they have become known as the plankton of the air.

Observers have noted that some of the insects fly backwards. These insects, which are actually being carried, flap their wings only enough to keep aloft during the trip. As the weather systems weaken, these passengers, like rain, are dumped rather unceremoniously from the sky.

Some of our common insect pests are weather-front hitchhikers. Many aphids, sometimes known as plant lice, come into the Midwest astride a storm. Leafhoppers, including the potato leafhopper, a pest of alfalfa, also arrive on the wings of a storm.

While aphids and leafhoppers are rather small insects, some larger ones also pick up a ride on the southern zephyr. One major pest of gardens and field crops is the black cutworm, which in the adult stage is a moth nearly an inch in length. This moth isn't at all bashful about spending the winter in the sunny South and then taking the first spring weather front to points further north.

This year, as she has done in the past, Mother Nature will most likely bestow on us one of her famous spring downpours. When she does, some will be tempted to say that it is raining cats and dogs. The truth of the matter is that it's likely raining aphids, leafhoppers and cutworms!

Comforting thought, isn't it?

Surviving Winter

1. What are two methods insects use to survive as a population in an Indiana winter?
2. Where does the monarch butterfly acquire its bitter taste?
3. Where does the monarch butterfly overwinter?
4. Name two insects that ride the spring winds into Midwestern states.

CHAPTER 16

Insect Societies

The Swarm

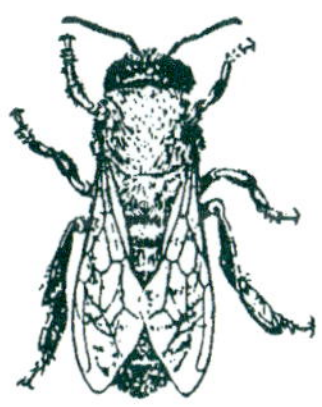
worker

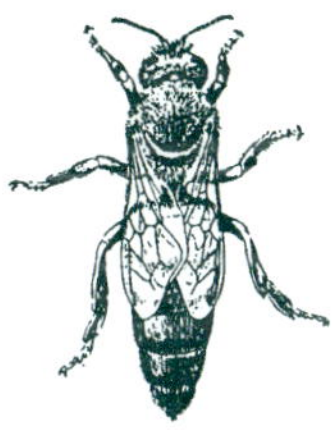
queen

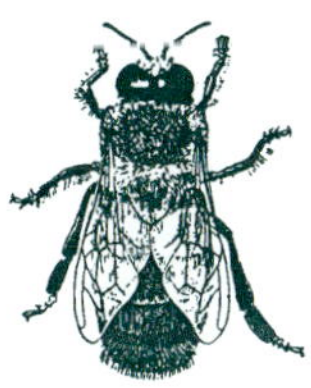
drone

The miniature denizens of the alien society spew forth from the honeycombed labyrinth of their darkened domicile. Armed warriors, in blind obedience to some prehistoric imperative, swarm about their deposed leader. Humans, mere mortals, tremble at the sight and sound of the frightening frenzy.

An advertisement for the latest horror movie? No, just one of the rites of spring—the swarming of honey bees. This process, which occurs primarily in the months of May and June, allows honey bees to establish new colonies.

The honey bee swarm is one of the true wonders of nature. First, a new queen is produced in the colony. The presence of the young queen causes the old queen and her loyal workers to leave. Before they leave, the workers fill their crops with honey—the insect equivalent to packing a picnic lunch.

The first flight of the swarm normally takes it to some temporary resting place. The bees cluster around the queen while scout bees search for a permanent abode. The cluster may be on a fence post, a tree limb or a street light. If the swarm happens to land in a populated area, it is sure to create a sensation. The fear of being stung by a swarm of bees is, however, mostly unfounded. Swarming bees are quite docile and seldom sting.

The amount of time the swarm spends at its first landing site is normally limited. Consequently, the best action when confronted with a cluster of bees is to do nothing; they are not dangerous and will soon leave. Once the scout bees have found a permanent site, the cluster of bees takes wing, and the queen is escorted to the new location. At the new site, the bees immediately get down to business and set up housekeeping.

Beekeepers collect swarms and put them in hives as a method of increasing the number of colonies in their apiaries. It's easy to do. Once the queen is in the hive, the dutiful workers follow. However, for a new colony to be successful, it must get established in time to gather enough honey to sustain it over the winter. This idea is incorporated in an old folk saying relative to collecting a swarm of bees. One version is:

A swarm in May is worth a load of hay,
A swarm in June is worth a silver spoon,
A swarm in July, just let them go by!

If in the future you happen to encounter a swarm of bees, don't panic. Just take the opportunity to enjoy one of the natural wonders of the world. Seeing is bee-lieving!

The Case of the Winter Wasp

Everybody knows that winter is a time when insects are scarce. You see, winter is that season when temperatures turn downward. Cold temperatures and insects don't mix.

Being cold-blooded, the insects can't function when it's cold. We still have our housemates, like cockroaches, fleas and flies, that have come to live in our homes and on our pets. These do quite well during the winter months! But most insects disappear.

Enter the winter wasp. The wasp, that armed warrior of the insect world and flying, poison-loaded hypodermic needle. The wasp, that uninvited purveyor of terror for every man, woman and child within striking distance.

Pollen-Packing Mamas of the Insect World

Pollination is the process by which most plants reproduce. Pollination results in seeds that germinate into new plants.

It is not a surprise to most people that pollen is a key ingredient in the pollination process. Pollen is the package that moves between parts of the plant or between plants for the purpose of fertilization. Sometimes plants are self-pollinating. Other times the wind carries the pollen from plant to plant. Still other plants depend on insects to transport the pollen. These plants generally have showy flowers to attract the potential pollen-packers.

Some insects that carry pollen are especially well equipped for the task. Most are covered with a heavy coat of hairs. Insects, like the bumble bees, are fuzzy in appearance. The hairs on these insects have a large number of hooks and teeth. When the insect goes into the flower, either looking for pollen or nectar, the hairs become covered with pollen.

What happens after the pollen-covered insect emerges from the flower is a fascinating process. It is a grooming miracle, using brush and comb, all while the insect is in flight.

On the foot of the hind legs of bees are pollen brushes. So while the bee flies to the next flower, it uses the brushes on its hind legs to brush its hairy body. In the process, the pollen is removed from the hair and collects in the brush.

The grooming bee then removes the pollen from the brush by using a comb that is on the lower end of its leg. This, of course, requires that both legs come into play. The comb from the left leg removes the pollen from the right leg and vice versa.

Once the pollen is in the comb of the leg, it is packed into what is called the pollen basket. This basket is just a hollow spot, ringed by two rows of spines on the inside of the back leg.

The middle legs of the bee are used to pack the pollen into the basket. The bee sometimes adds a little honey to the pollen to help produce the solid mass that the bee will carry back to the hive. Once in the hive, the pollen packets are removed from the legs of the field bee and placed into the pollen cells in the comb. There, it is packed into the cells by hive bees where it will remain until it is needed as food.

Not all bees collect pollen with their legs. Some bees have an abdominal brush. In these bees, such as the leaf-cutter bees, the pollen is carried to the nest in the brush. There, it is dusted off so that it can be concentrated for storage.

Other pollen-carrying bees just swallow the pollen that is collected on their bodies. It is carried to the nest in the crop along with the nectar. When these bees get home, they just spit out the gut contents and it is stored for later use.

Humans have come to recognize that pollen might have benefits for their health. Some of us who suffer from hay fever, due to air-borne pollen, might beg to differ, though. Nonetheless, there is a market for pollen.

Some beekeepers put pollen traps on their bee hives. As the field bees carrying pollen enter the hive, the trap causes the pollen packets to drop from the bees. The beekeeper then collects the pollen so that it can be packaged and sold at health food stores.

So maybe the sales slogan for the pollen might be something like this: Brushed and combed from the hairs of knees of pollen-packing mama bees!

Honey Bees Not Native to North America

Honey bees are among the most recognizable and beneficial of the insects that live in North America. But these insects are not even native to the Americas. Like most of the livestock associated with American farms, honey bees were imported by European settlers.

Prior to the arrival of the Old World settlers, honey bees were unknown to Native Americans. In fact, several early American writers, including Thomas Jefferson, reported that honey bees were called "white man's flies." The name was a recognition that the appearance of honey bees in America was associated with the arrival of the Europeans.

There was a close association between the westward migration of Europeans and the establishment of wild colonies of honey bees. Native Americans were said to have noticed that shortly after colonies of honey bees were discovered, white settlers would not be far behind.

So when did the first colonies of honey bees arrive in the New World? These bees probably came from England and arrived in Virginia in 1622. By 1639 colonies of honey bees were found throughout the woods in Massachusetts. Some of the colonists who arrived at Plymouth likely brought bees, as well as sheep, cows and chickens on the trip across the Atlantic.

Once the bees were introduced, they, like other insects, were able to increase their range by moving into new territory. Honey bees increase colony numbers by swarming. Swarms are able to fly several miles to establish a new colony.

Such migrating swarms brought honey bees to Connecticut and Pennsylvania by the mid 1650s. Honey bees had swarmed their way into Michigan by 1776 and Missouri, Indiana, Iowa and Illinois by 1800. In the next 20 years or so, bees had made their way to Arkansas, Oklahoma and Texas, as well as Wisconsin.

Further westward migration of the honey bee was slow. In 1843 it was reported that there were no honey bees beyond Kansas. However, Mormons arrived in Utah, and the first bees were taken there on the back of a wagon in 1848. So successful was this introduction, it was reported that a considerable amount of honey was being made in the southern counties of Utah. By 1852 the swarms had reached Nevada.

Bees were finally introduced into the Pacific Coast states by using a sea route along the East Coast and crossing Panama, before using the Pacific Ocean for the final part of the journey. It was in 1853 that botanist C. A. Shelton used this route to introduce the first honey bees into California. Only enough bees from 12 colonies survived to establish one colony, but it was enough to allow history to credit him with starting the honey bee industry in the golden state.

Transporting colonies of bees either by sea or land in the 1700s and 1800s was not easy. The sea voyage from England lasted six to eight weeks, and it was not easy to keep bees alive for that length of time while confined. Many of the attempts to transport bees were unsuccessful as many stories relate.

For once in our history, the introduction of a foreign insect has a happy ending. After all, honey bees are a very important part of agriculture in this country, and we really can't do without them. Even if they do sting us once in a while!

Honey Bees According to Aristotle

Aristotle, the ancient Greek philosopher, seems to have known something about everything. In his writings, he touched on all kinds of topics, even insects. In fact, his writings include a detailed description of honey bees.

Some authorities suggest that the honey bee information attributed to Aristotle was written by someone else. This person has become known as the "Aristotelian Bee-Master." Regardless of the actual authorship, the information included is remarkably accurate. Especially considering that it was compiled some 350 years B.C.

In Aristotle's works, the honey bees are lumped together with other insects that build a honey comb. This group is further divided into gregarious and solitary forms. Amongst the gregarious types are the bee, the king-bee, the drone-bee, the annual wasp, the hornet and the ground wasp.

The bee, the king-bee and the drone-bee are what we know today as the three castes of the honey bee. The bee is what we call the worker today. The king-bee is in actuality the queen. This misunderstanding of the sex of the ruler of the colony is the biggest mistake included in the writings about the honey bee.

It is interesting to note that reproduction of the honey bee was poorly understood in Aristotle's days. His writings on the history of animals included some hypotheses. First, there was the belief that bees do not give birth to young but fetch their young from flowers.

Even the type of flower was in question. Some thought the olive, others the reed or the callyntrum. Some people apparently believed that the drones came from the flowers, but the working bees were engendered by the rulers of the hive.

In reference to the rulers, Aristotle wrote that "they are called by some the 'mothers' from an idea that they bear or generate the bees." As proof of this theory, it was pointed out that the brood of drones appears even when there is no ruler bee in the hive. This is an accurate observation. As we know today, unfertilized female workers can lay eggs in the absence of a queen, but the eggs develop into drones.

As for the development of the bees in the immature stages, Aristotle was right on! He stated that the ordinary bee is generated in the cells of the comb. The ruler-bees, on the other hand, come from larger cells suspended on the bottom of the comb.

It is further stated that the young larvae are fed in the cells where they reside. When completely fed, the cell is sealed up with wax, which the new bee must chew through when it emerges.

Aristotle knew that bees obtained material from flowers. But just what was in question. He stated that the honeycomb is made from flowers. The materials for the wax comes from the gum of trees. It was suggested that honey actually is deposited from the atmosphere. As proof of this, Aristotle offered two facts. Occasionally, bees fill the combs with honey in a day or two. Also, in the fall, flowers are found but very little honey is produced; therefore, it couldn't be coming from flowers since they are abundant. Today, we know that fall flowers are not good producers of nectar and, as a result, bees do not produce much honey from them!

What people of Aristotle's time did know is that the bee carried material around on its legs and vomited the honey into the cell in the comb. Yes, even in Aristotle's day, the young boys at the dinner table could gross out their mothers by demanding some "bee vomit" to spread over their breakfast toast!

Summertime Abuzz with Bees and Wasps

The good ol' summertime means lawn mowing, baseball games, picnics in the park, family vacations and encounters with insects. Human and insect interaction during the hazy, crazy days of summer can be pleasant experiences—a butterfly nectaring at a flower or the flashing lights of hundreds of fireflies over a misty nighttime meadow.

Many times, though, summertime encounters with insects are of a kind that we like to avoid, like a bite from a mosquito or a sting from a bee or wasp. Yes, the summer buzzing of insects can mean an unpleasant situation.

But wait, not all insects that buzz will sting. Insects that intentionally sting are bees, wasps and some ants. A few caterpillars have stinging hairs and are to be avoided, but you have to touch them to get stung.

To be sure, all female bees and wasps possess stingers. But not all bees and wasps will sting humans. Some only use their stingers to paralyze food for their young.

Generally, the bees and wasps that are called social insects will sting. These insects live in a colony, which can include hundreds of individuals. Well-known stinging social bees are bumble bees and honey bees.

Social wasps that sting include the so-called paper wasps. These wasps construct their nest material out of chewed wood or paper. One species of paper wasp makes its open nest under the eves of buildings. Another paper wasp is known as the bald-faced hornet. This hornet constructs a crepe-paper globe nest that hangs from a tree limb. Yellow jackets are also hornets, but their nest is constructed underground.

These social bees and wasps use their stingers to defend their nest. So, if you get too close to the nest, the inhabitants may protect it by stinging you, the intruder. Many a young farm boy has provoked a wasp or bee colony into defensive behavior and received multiple stings for his trouble!

Not all bees and wasps are social insects. In fact, most species of bees and wasps live solitary lives. The females of these species raise their offspring alone. Generally, the solitary bees and wasps use their stingers only to paralyze their prey. While they can sting other things, it is rare that they do so.

Some solitary bees make their nest by mining in plant stems or wood. The carpenter bees build their nest in wood, sometimes in the wood that is the structure of our homes. Some leaf-cutter bees mine the pith out of plant stems for their nest. All of these bees feed their offspring pollen and nectar.

A few species of solitary wasps also feed pollen to their young. Most, however, provide their larvae insects or spiders as food. Mud daubers fit into this category.

Mud daubers get their name because they construct their nest of wet soil. The nest of some species looks like a batch of mud thrown against the side of a building. Other species are known as organ-pipe mud daubers because the nest resembles the pipes of a church organ. In both cases, the adult female wasp provides paralyzed spiders in the cell of a nest as food for the offspring. As many as 15 spiders might be placed in a mud dauber cell before an egg is laid.

Most of the wasps that feed on insects and spiders burrow in the ground when nesting. One of these wasps is known as a tarantula hawk because it provides tarantulas for its kids' dinner. This, as most people would guess, is a large wasp.

Another large wasp is the cicada killer. As the name suggests, this wasp catches and places a cicada in a burrow as food for the offspring. Fortunately, the cicada killer is a very calm insect and hardly ever stings humans. It should wear a sign that reads, "I will sting for food."

Everybody knows that wasps aren't to be about in winter! Except for a few wasps. For example, there was a wasp in our kitchen the day after Christmas. My daughter vociferously called it to my attention. And what winter churchgoer hasn't been a witness to the quiet hysteria of the congregation that is created by the menacing midwinter flight of a wasp. Such wasps always seem to gain attention by losing a significant amount of altitude right over the "Amen" pews.

This wasp, out of place, is a quirk of wasp biology brought about by heated buildings, churches, houses and the like. In the fall, with the approach of winter, many female wasps mate and head to sheltered locations to hibernate for the winter. A typical natural location for hibernation would be in the leaf litter of the forest floor, where a wasp would settle down for a long winter's nap.

With the coming of spring, the rays of the sun warm the forest floor. The warmer temperatures signal to the slumbering wasp that it's time to venture from the hibernation location. She heads on her way with the mission of beginning a new nest for the coming season.

But in permanent buildings, such as houses, churches and schools, the lady-in-waiting wasp has a new place to spend the winter. However, buildings, which are unnaturally warm to insects during the cold months, cause some problems. The wasp that has crawled into the inner crevices of an attic, eave or windowsill to slumber is exposed to the heat from the interior of the building. Consequently, she does the perfectly normal thing for a wasp—begins to move around in preparation for nest-building and egg-laying. That results in the flight of a winter wasp.

The strafing run above the congregation during the sermon on a bright Sunday in February is sure to grab attention. Don't think of that brazen, aerial daredevil as a wasp out of place. Think of her as the first indicator of spring!

Clay Masons

Clay is widely used as a raw material for home construction by insects and people. Adobe and brick

The Yellowjackets Are Coming

Fall—the harvest season. That time of the year when pumpkins lie in golden contrast to the fading green vines of their nurturing plants. When sun-ripened apples are turned into sweet mush under the relentless crunch of a cider mill. When the quiet serenity of a backwoods picnic or suburban cookout is suddenly broken with a blood-curdling scream—yellowjackets!!

Yes, folks everywhere slightly modify Paul Revere's now-famous warning of Revolutionary War times. "The yellowjackets are coming, the yellowjackets are coming" is an oft-repeated warning of close encounters of the insect kind.

Yellowjackets are common, ground-nesting, social wasps. In the spring, a mated female emerges from her overwintering site and establishes an underground nest in an abandoned rodent burrow. There she begins the task of rearing young wasps. Early growth of the colony is slow, because the queen does all of the work. However, by midsummer other wasps have been produced. These workers take over the duties of food finding, larval care and defense of the nest. By late fall, some yellowjacket nests may contain as many as 3,000 workers.

Immature yellowjackets are fed meat, mostly in the form of arthropods captured by workers. In the fall, when prey species become scarce, the yellowjackets become scavengers. They can be found around garbage cans and are frequent visitors to picnics. That, of course, causes great concern to humans, who are reluctant to share the feast with six-legged visitors, especially those armed with a stinger.

In general, yellowjackets do not sting except in defense of their nest or when physically abused. Therefore, remaining calm is the best policy when faced with a yellowjacket intent on tearing shreds of meat from your bologna sandwich or skating on the ice cube in your cola. Aggressive behavior normally prompts a like response on the part of the yellowjacket, and in most cases, the insect will be victorious.

If yellowjacket nests in lawns are a continual problem, they can normally be eliminated by finding the hole to the nest and treating with an insecticide during the night. However, such activities can be hazardous to your health, because yellowjackets will not stand idly by while their home is being destroyed.

In most instances, folks are well-advised to leave the nests alone. Yellowjackets are beneficial because of the insects they destroy, and winter will solve the problem for another year. Besides, who wants to be guilty of stirring up a hornet's nest?

houses have sheltered humans for centuries. Insects, however, were using clay for homes long before humans abandoned cave dwellings.

Among the finest of the insect clay masons are the wasps. Two of the most common are the potter wasps and the mud daubers.

Potter wasps, also called mason wasps, are common insects. They construct their nests of mud and attach these jug-like homes to twigs. The home builder provisions the nest with caterpillars. From an egg attached to the mud home by a string, the newly-hatched larvae descend to feed on the food that mother supplied.

Mud daubers are aptly named. Their nests are constructed by daubing mud in a somewhat organized fashion in reasonably protected locations. Wasps of this group can frequently be found along the edges of ponds or streams. Here the wasps scoop up the mud in their mouths for transporting to the nesting site. These wasps are frequently blackish-blue in color.

One of the most recognizable of the mud daubers is the pipe-organ mud dauber. The nests of this wasp are laid end to end in tubes. Several tubes are constructed side by side in a pattern similar to the pipes of a fine church organ. Each cell is provisioned with a spider as a food resource for the young wasp. When the wasp has completed development, it chews a hole through the side of the tube.

Some mud daubers use any available facility for constructing a home. In fact, old-fashioned outhouses provide ideal nesting sites! Anyone who has had the occasion to use an outhouse in the summer is well aware of that fact.

The composer of the song "Ode to the Little Brown Shack Out Back" noted the presence of wasps in such a building. However, the wasp was misidentified in the phrase "to the yellowjackets' drone." That non-human singer was certainly a mud dauber. And some people would say you haven't lived unless you have used an outhouse—at least used one while being serenaded by a female mud dauber as she fashions her nest.

Of course, a mud nest under construction requires mud. Mud-laden mud daubers frequently fly from the quarter moon in the door to the nest site. Such a flight pattern is sometimes over the best "seat" in the house. For the human occupant of the outhouse, such flights can be exciting. A little too exciting! Some folks have been known to leave the facility rather abruptly when faced with a low-flying wasp.

That's the way it is with some humans. They just don't appreciate insects such as mud daubers—at least when they are sitting in an outhouse reading the Sears and Roebuck catalog.

Warrior Ants on the March

It is the stuff horror movies are made of. Hordes of seemingly mindless individuals terrorizing every living thing in their path. Unafraid of humans or any technology that the human mind can conceive, these creatures use fearsome jaws to wreak havoc on all they encounter.

"They" are warrior ants, sometimes called army or driver ants. No wild-eyed scriptwriter has to dream up such a movie tale. It already exists!

Warrior ants live in the semitropical regions of South America and Africa, but unlike most other ants, they do not have a permanent home. Warrior ants form bivouacs from which they raid surrounding areas in search of plunder. Or they go on marches during which it is said they consume all of the animal refuse in their way. During such frenzied marches, they will not hesitate to attack all kinds of vertebrates, including human beings.

Many are the tales of encounters with the warrior ants. Dr. Albert Schweitzer had to conquer the warrior ants, among other things, as he developed his hospital in Africa. Of particular interest to Schweitzer was the chicken house. At the onset of a warrior ant invasion, Schweitzer would jump out of bed and run to open the chicken house door. According to the famous doctor, "Shut in, they would inevitably be the prey of the ants, which would creep into their mouths and nostrils until they are suffocated, and then devour them, so that in a short time nothing is left but their white bones."

Another well-known doctor who devoted his life to Africa also recorded encounters of the unpleasant kind with warrior ants. This doctor was none other than the one who prompted the journalist Sir Henry Stanley to exclaim upon finding him living on Lake Tanganyika, "Dr. Livingstone, I presume."

Dr. David Livingstone recorded in his diary that the camp "suffered a furious attack at midnight from driver ants." In an attempt to keep the ants away, Livingstone's men lighted grass fires. In his last diary entry for the day, he wrote: "We put hot ashes on the defiant hordes." Apparently to no avail because the good doctor noted that he had suffered bites so numerous that he resembled a person who had smallpox.

In some parts of Africa, criminals were punished a century or more ago by binding their hands and feet and laying them in the path of an ant army. Francois Coillard, a missionary in Africa, reported that under such circumstances, "in a surprisingly short time the writhing victim will have been changed into a skeleton of clean and polished bones that will make the trained anatomist envious."

Over time, such observations of warrior ants have no doubt helped give credence to mother's admonitions to fidgety children: "You're wiggling around like you have ants in your pants!"

Insect Societies

1. What is the biological purpose of swarming in honey bee colonies?
2. Swarms of honey bees are normally docile. One reason is an action they take before leaving the old hive. What is the action?
3. Is it the old queen or the young queen that leads a swarm of bees?
4. Warrior ants are found in what kind of climate?
5. What is the larval food of yellowjackets?
6. Why might yellowjackets be considered beneficial insects?
7. What is a common overwintering form of many social wasps?
8. With what do pipe-organ mud daubers provision their nests?

CHAPTER 17

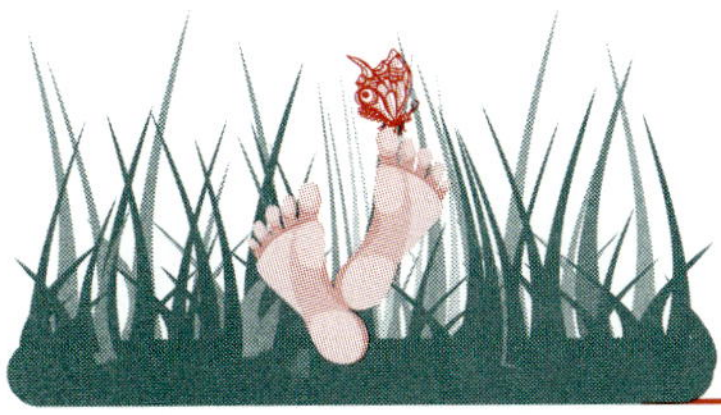

Protection from Predators

Insect Eaters

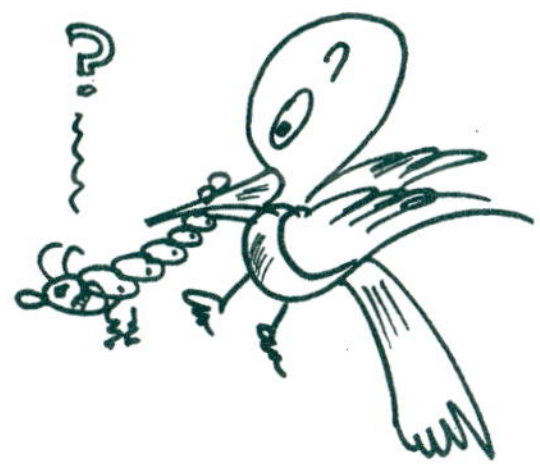

Why are there insects? It's a question as old as the ages. Odgen Nash alluded to it in his poem "The Fly:"

> *God in his wisdom made the fly, and then forgot to tell us why.*

Nash is not alone. Almost everyone at one time or another has wondered why insects exist.

One important ecological service insects provide is nutrition for other animals. There are a lot of insect eaters out there. Some even have names that reflect their culinary habits.

Most people are aware of anteaters. These animals spend their entire lives dining on ants. And there's a bird called a flycatcher. Its name alone suggests its favorite food—flying insects.

Many different animals use insects as a food source even if their names don't reflect it. Frogs and toads have sticky tongues to capture flying insects. Snakes and salamanders also will dine upon insects.

Many aquatic creatures feed upon insects as well. For that reason fishermen use artificial bait modeled after insects. These models are appropriately called "flies." Fishing flies come in many shapes and colors, but you can bet they are designed to entice the fish into thinking it is about to eat a tasty insect.

Mammals also eat insects. Bears eat grubs that they dig out of rotten stumps. Skunks also eat insects. A grub-infested lawn is a great place for a skunk to dig up a meal. This process leaves holes in the lawn, much to the disgust of the homeowner. Even dogs and cats will chow down on an insect or two when given the chance.

Bats, the only flying mammals, depend upon insects for their entire food supply. Bats fly at night and use echolocation to navigate and catch night-flying insects, sort of a meal on the wing.

Many birds consume insects. Woodpeckers peck wood looking for larval insects. Wrens are great insect eaters. Many blackbirds also eat insects. A flock of blackbirds feeding on the lawn in the spring could be a sign of an insect infestation. That domestic fowl, the chicken, also delights in a good insect meal. Years ago, chickens running in the orchard were considered beneficial because they ate the insects that were overwintering in the fallen fruit.

Why are there insects? They keep a lot of animals from going to bed hungry!

Deceitful Buzzers

Buzzing is universally regarded as a warning sound. For example, buzzers remind us to fasten our seat belts. Buzzers in smoke detectors alert us that our houses are on fire or that the roast in the oven is slightly overdone. Elevators, airplane guidance systems and even computers buzz when things are amiss.

Buzzing as a warning signal was perfected by insects, especially the bees. The sound is produced by wing vibration as the insect flies. Most animals respond negatively to the buzzing of bees, and for good reason. Bees are equipped with a stinger they can use for defensive purposes. However, in most encounters with other animals, bees do not need to unsheath their stinger. Their buzz alone wards off most sources of potential danger.

So effective is the buzz of bees as a protective device, that other insects have adopted it for their own use. Many flies, because their wing-beat frequencies are similar to that of the bees, sound like bees. The sound alone usually provides some pro-

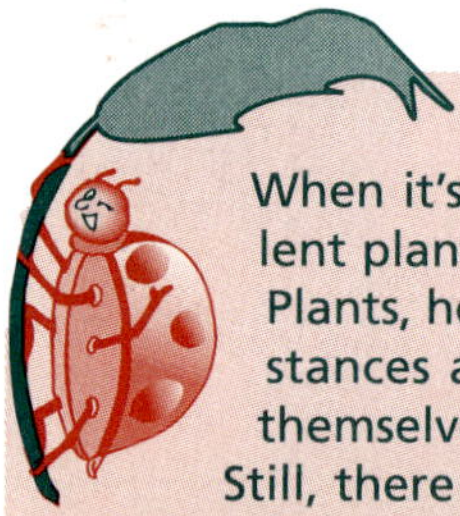

Turn About Is Fair Play

When it's time to dine, insects choose their favorite foods: tasty plant leaves, stems and roots; succulent plant sap; and delectable plant flowers and fruits.

Plants, however, are not eager to be devoured by insects. Many plants have spines and hairs, sticky substances and even bad tastes to thwart hungry mandibles. Yet in spite of their attempts to protect themselves, most plants become the victims of insects.

Still, there are a few plants that have turned the dining tables on insects. These plants, called insectivorous plants, trap insects and make a meal of them.

The best known of the insect-eating plants is the pitcher plant, a native of North America. Its leaves and petioles are modified to form hollow "pitchers." Glands on the inside of the pitchers secrete a substance that attracts insects. Insects trying to reach this substance fall into a liquid at the bottom of the pitcher and are digested by the plant.

The sundew plant has leaves with numerous sticky hairs. Insects that get trapped in the hairs are digested in a fluid secreted by the leaves.

Another insect-eating plant is aptly named, the Venus's-flytrap. This plant has leaves that end in two lobes. The lobes are bordered by a series of spines. The plant just sits around with the lobes expanded waiting for a meal to come by. When some unsuspecting fly wanders into the center of the expanded lobes, the lobes will spring shut like the jaws of a steel trap. The fly cannot escape this plant. It is doomed to be digested by a secretion poured from glands on the surface of the leaf.

For these plants, eating insects is more than a sinister method of getting even. Insect meals provide insectivorous plants with nitrogen. Most plants that eat insects live in wet areas where nitrogen is limited in the soil, so they need to supplement their nutrient uptake.

It's ironic that for a few plants a tasty insect not only provides a fine meal, but also gives them gas—nitrogen gas.

tection for these imitators. However, when their color pattern matches that of bees, only an entomologist can tell the difference at first glance.

Some flies even mimic the aggressive behavior of disturbed bees. One group of flies that sounds, looks and acts like bees even claim their mentor's name—bee flies.

Not only flies have benefited from sounding like buzzing bees. One beetle, the green June beetle, is especially adept at convincing folks it is a dangerous insect. This is a large, somewhat flattened, green beetle with yellow body margins. As an adult, the green June beetle feeds on the foliage of fruit trees and on the ripening fruit. When these beetles appear in large numbers and swarm around during their feeding activities, many people have been convinced that they have encountered a swarm of killer bees. The beetle is perfectly harmless, but most folks and other animals don't stay around long enough to find out.

Some might suggest insects that mimic the sound of bees are guilty of fraud or false advertising. But think of it this way, a little benevolent deceit can go a long way toward preventing encounters of the worst kind. At least from the insect's point of view!

Stingers and Nettlers

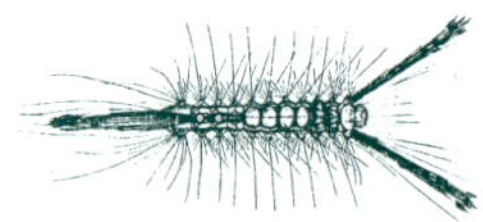

This world is full of stingers and nettlers. Stingers like scorpions, bees, hornets and wasps actively inject poison into their targets. They deliberately and aggressively pursue their victims. The insect stingers even sound dangerous, buzzing menacingly before they sting.

Nettlers are the mild-mannered poison peddlers of the world. Nettlers are passive stingers. They include stinging plants; some are appropriately called nettles, jellyfish and some caterpillars. Nettlers never look for trouble, they simply wait for the victim to contact them. So if you get too close, you are likely to regret it.

Stinging and nettling are methods of protection for these creatures. Stinging allows insects to defend their homes as well as their lives. Both stingers and nettlers use their poison-peddling prowess to avoid becoming a meal for some hungry insect or plant eater.

Even some fuzzy caterpillars are poison warfare experts. These insects possess poisonous hairs which are typically shorter than other caterpillar hairs. The hairs are barbed and connect to a poison gland at their base. When the poisonous hair is contacted, the poison is released, and the victim experiences a stinging sensation.

Some of the nettling caterpillars are brightly marked. One such is the saddle-back, a pale green insect with a brown and white design across its back that resembles a saddle. Other nettling caterpillars include some of the tussock moths, the pus moth and the buck moth.

Years ago, witch doctors of primitive tribes recognized that some caterpillars had natural stinging arrows on their bodies. These witch doctors frequently employed the irritating power of the poisonous hairs of these insects in their witchcraft.

Anyone who has had an encounter with a nettling caterpillar can tell you that they, like fire, are beautiful to behold but dangerous to touch. And that is the way the insect would like to keep it.

Protection from Predators

1. How do some flies use the sound of their wings for protection?
2. What type of butterfly gathers to rest on "butterfly trees?"
3. What are cardiac glycosides?

Beautiful but Bitter

It's yellow, black and white striped in its youth, has 14 leg-like appendages, fearsome mandibles and a hard-shell head. It feeds on a plant that contains chemicals used as a medicine. Later, as an adult, it's brown with black stripes, and spends the winter in a remote mountainous region of Mexico. Could it be an invader from outer space or a creature from the black lagoon?

Well, actually, it's the monarch butterfly! One of our best known insects. The bright, striped colors of the larva warn potential predators that it's distasteful.

The bitter taste of the larva comes from the juice of the milkweed plant that it uses for food. The milkweed plant contains bitter substances called **cardiac glycosides**, which are named for their use as a human heart medicine. These compounds benefit the milkweed plant by acting as antifeeding compounds for most animals, including insects.

The monarch, however, has developed the ability to utilize the plant in spite of the presence of the bitter chemicals. The larva stores the toxic compounds in its body. This makes the larva bad-tasting, an attribute that is transferred to the adult following pupation.

Many birds are predators of insects, but they can't tolerate the bitter taste of the monarch. To advertise their bad taste, the monarch, in both the adult and larval stages, wears bright colors. So effective is the scheme that the Viceroy butterfly, a good-tasting insect which resembles the monarch in color and pattern, also is protected.

The monarch, actually a tropical butterfly, has adapted to more northern climates by leaving the winter behind. Each fall monarchs leisurely flutter toward overwintering areas in the mountains of Mexico.

Along the way, great flocks of these insects sometimes gather on trees to rest overnight. These resting sites, called butterfly trees, are one of nature's truly memorable sights.

The monarch butterfly is indeed a biological miracle. These fragile insects wing over 1,000 miles to a place they've never been. Come spring they head north to lay eggs on plants too toxic for most animals to eat.

Once egg laying is complete the butterflies die, leaving the younger generation to carry on. Each succeeding generation moves northward to lay eggs until their genetic code unerringly indicates that it's time to turn around.

The butterflies begin the long and treacherous southern journey. They make the trip not so much for themselves, but to ensure survival of future generations of the "beautiful but bitter" monarch.

Birds Really Go for Bug Chow

Animals that feed on insects are known as insectivores. There are many kinds of insectivorous birds. In fact, the *Peterson Field Guide to Eastern Birds* lists 64 families of birds, and 37 of these families contain species that feed on insects. That does not count birds that dine on other arthropods. For example, some ocean birds feed entirely on insect-related crustaceans like shrimp.

Of the birds that will make a meal of insects, some feed on insects when available, while others feed exclusively on insects. As their names suggest, the flycatchers and gnatcatchers are part of the latter group. The flycatchers include such familiar birds as the kingbirds and the phoebes. These birds sit on a fence or branch to sally forth and snap up flying insects. The kingbirds are known to even feed on honeybees and are, therefore, not the favorite birds of beekeepers.

The goatsuckers, including the whippoorwill and the common nighthawk, use only insect food collected during the nighttime hours. The nighthawk gets its common name from its habit of diving and swooping in search of night-flying insects.

Other birds that depend on insect food include the familiar swallows and swifts. As these birds fly around during the day, they catch small insects. It is the reason swallows fly in front of a tractor that is mowing hay, since the mowing operation flushes insects from the vegetation and provides a dining opportunity for the birds.

Vireos, warblers, wrens and finches also feed on insects, and many are exclusive insect feeders. One only has to sit and watch a pair of wrens feeding their young to realize that many insects are consumed by these birds during the course of the summer. Unlike most other insect-eating birds, wrens will also feed on spiders.

Most birds that depend entirely on insect food are summer residents of temperate regions. As winter approaches, their cold-blooded insect food disappears and these insect feeders are forced to move south, some even to the southern hemisphere, in order to find food. For most of us, the return of swallows and wrens to our lawns and barns is a joy. It is a sure sign that winter is over, but it also signals that the insect food is back!

One totally insectivorous group of birds does spend the winter in North America. Since the insect food of woodpeckers is under the bark of trees, it is there for the taking in summer and winter.

Other birds that eat some insects include kingfishers and hummingbirds. Turkeys, grouse, quail, pigeons and doves also will not turn down a tasty insect as food. Even some of our least favorite birds, the starlings and house sparrows, consume many insects during the summer months. To some gardeners it is nice to know that even the house sparrow has at least one redeeming characteristic!

CHAPTER 18

Population Dynamics

A single insect, unless it happens to be a mosquito buzzing in your ear or a bee about to sting, is generally of little consequence to humans. It is populations of insects that really get our attention. Consequently most studies of insects are based on populations, not individuals.

Insects have a tremendous biological potential based on their rates of reproduction and short generation time. This potential is generally characterized as a **logarithmic rate of growth**, a J-shaped curve on a graph that relates population to time (Figure 18.1). This rate of growth doesn't continue unbridled in nature, otherwise we would be up to certain portions of our anatomy in insects. Nature applies brakes to logarithmic growth. Those brakes come into play as the population approaches the **carrying capacity** of the environment (depicted as "K" on many graphs) (Figure 18.3).

Some factors that limit a population as it approaches carrying capacity for the environment are lack of food or water, buildup of toxic wastes, weather, and predators and parasites. One aspect of insect control is to use the factors that limit populations to our advantage in reducing the number of insects we consider pests. For instance, removal of crumbs from behind the sink (limit food) will help keep cockroach populations down. We can also place insect infested dried-flower arrangements in the freezer for a day to remove the undesirable creatures.

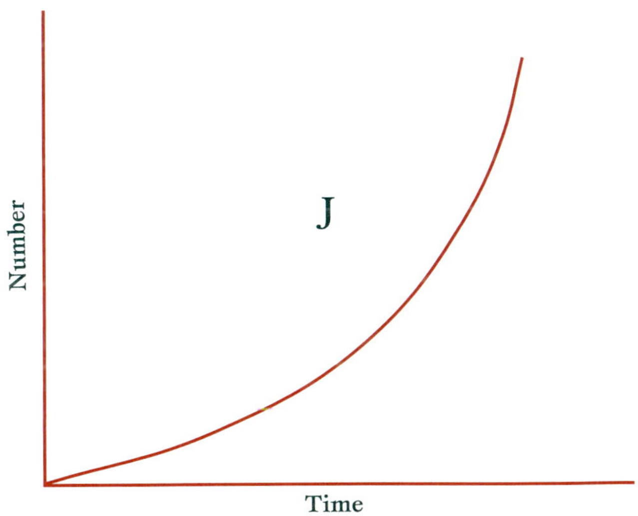

FIGURE 18.1. Theoretical population growth curve without restraints.

Since populations are the key to most insect situations, entomologists frequently try to monitor population buildup. This can be done by either counting the number of insects involved or sampling. Counting all the insects in an area is nearly impossible unless the area is very small (like a jar) so most entomologists estimate populations. Population estimates are done by sampling.

Insect sampling involves several techniques. One is to capture the insects and count them. This can be done with an insect net. Sometimes the traps depend on the insect to come to it. Such is the case with a light trap, a pheromone trap, or a sticky trap (something like fly paper). Of course any such method must be related to the number of insects that are really there. One method of calibrating a sampling method is called **capture-mark-release-recapture.** In this mark and release procedure, a given number of insects are captured, marked and released. When a sample is taken, the number of marked insects relative to the total captured can be used to estimate the total number of insects available in the area.

Another method of insect sampling is to count symptoms such as holes in leaves or signs such as frass or cast skins. Such methods are especially important in agricultural entomology where the damage and not the insect is the important issue.

It is sometimes important to predict the buildup of an insect population. The buildup can be influenced by a number of factors including the population limiting factors mentioned above. In addition,

Most people have heard of *Drosophila,* the pomace flies that have been studied so much by geneticists. These flies develop rapidly and, under ideal conditions, may produce 25 generations a year. Each female will lay up to 100 eggs, of which half will hatch into males and half into females. Now, suppose we started with a pair of these flies and allowed them to reproduce under ideal conditions for a year—with the original and each succeeding female laying 100 eggs before she dies, and each egg hatching and the young growing to maturity and reproducing again. The number of flies that would be produced in the twenty-fifth generation is fantastic (about 10^{41}); if the flies of this generation were packed tightly together, 1,000 to a cubic inch, they would form a ball of flies 96 million miles in diameter, or a ball extending nearly from the earth to the sun!

D. J. Borrer and D. M. DeLong

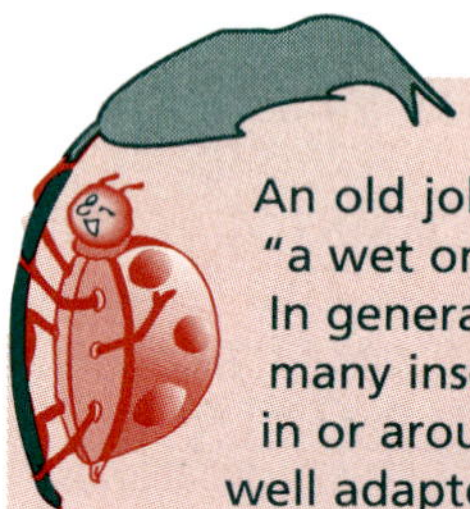

Rainy Days Bad for Hair and Bugs

An old joke asks: "What kind of leaf does a rabbit sit under during a rain?" The answer, of course, is "a wet one!" The same could be said of insects.

In general, insects are not well-equipped to survive falling rain and the water that results. To be sure, many insects benefit from rain. Especially those that live in water. Some insects spend their entire life in or around water. The names of water bugs, water striders and back swimmers suggest that they are well adapted to deal with water.

Other insects develop as immatures in water. For instance, that summer-time scourge the mosquito is a water dweller as a larva. The same is true of dragonflies, mayflies and caddisflies. Larvae of other insects, such as horse flies and deer flies, live in nearly saturated soil. Immatures of fireflies, called glowworms, live in damp soil along creeks and in swamps.

Most insects need moisture in their habitat but could do without the rainstorms on the plains, or in the forests for that matter. Millions of insects die as a result of rain and wind.

Insects not adapted to living in saturated soil will drown if the soil is water-logged for an extended period. Some insects that form their pupal chamber in the soil actually will waterproof the chamber. That works to keep moisture in during dry periods and out when the soil is saturated.

Any insect that lives exposed on plants can be harmed by a downpour. Many insects literally are knocked off their food plant by hard rains. Many times the insect will become stuck in the mud and die. Some caterpillars, even if they are not killed when knocked off the plant, are unable to find their way back to a food plant and die of starvation.

Newly hatched, small and soft insects can be smashed by a drop of rain. Very young insects frequently drown in water trapped in leaf axiles of plants. This can be of benefit when the insect is a pest. For instance, when young corn borer larvae drown behind the leaves of corn, it may eliminate the need to chemically control the pest.

Wet weather also has an impact on adult butterflies and moths. These insects with large scaled wings don't fare well in thunderstorms. Wind and rain can knock scales from their wings; it can tear the structure as well.

Butterflies do the best they can to take cover when a rainstorm approaches. Generally, they hang upside down under a leaf or a stem. With their wings folded and hanging down, little surface is exposed to the elements.

Other insects use structures to gain refuge from storms. Many wasps build their nests in barns and garages or under eves for that reason. Some adult insects move into shelters during storms. That is why fly populations increase on porches and in homes prior to and during a rain.

While severe storms are a hazard for humans, any hard rain is a problem for insects. For most insects, a rainstorm is a life-and-death matter, not just a bad hair day!

individual insects can leave (emigration) or come into the area (immigration) both of which would change the numbers. In fact with many pest insects, it is movement into a house, a field or even a country that is the beginning of the problem.

Since insects are cold blooded, their rate of population growth is tied to temperature. Consequently understanding temperature relationships can be useful in predicting a potential problem with insects. One method used to predict insect development has been with heat units. A **heat unit** is based on a threshold temperature (the lowest temperature where development will occur) for a specific insect. As temperatures exceed the specific threshold, heat units accrue and can be used to predict when the insect will reach various stages in development. Such a prediction can be used to schedule scouting or control activities.

Insect Population Dynamics

Why study insect populations?

1. **From human perspective.** It is insect populations, not single insects, that compete with humans.
2. **From ecological perspective.** Insect populations influence the environment in many ways. For example an insect population might destroy all of the plants on which it is feeding.
3. **From an insect perspective.** The insect populations might change because of external factors such as the environment.

Insect population study boils down to numbers of insects and rate of growth of the population. (The same is true for population study of any other animals including humans.)

Population growth is recorded as numbers of individuals over time. In theory, if such growth is not restrained it could be described mathematically as logarithmic (also called exponential) and would form a J shape when graphed. See Figures 18.1 and 18.2.

Populations cannot grow unrestrained forever. Most populations exhibit an S-shaped growth curve. (Figure 18.3) These populations level off at a point known as the **environmental carrying capacity**. The carrying capacity is generally called "K" on population graphs.

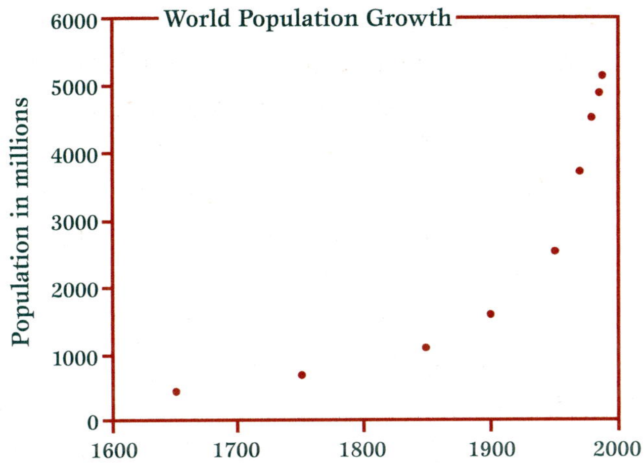

FIGURE 18.2. *World Growth Curve of the Human Population.*

Populations sometimes don't approach the carrying capacity of the environment because of what are known as **density independent factors.** Weather is one such factor. Generally, if some aspect of weather such as rain or temperature kills insects the percentage killed is the same, therefore it is said to be independent of density.

At other times, populations stop growing or decrease because of **density dependent factors.** These factors come into play when the populations of the insect (or other animal) approach some level. These factors include the following:

A. Predators, parasites and diseases.
B. Competition between different species such as between cows and grasshoppers for grass to eat. This is **interspecific competition**.

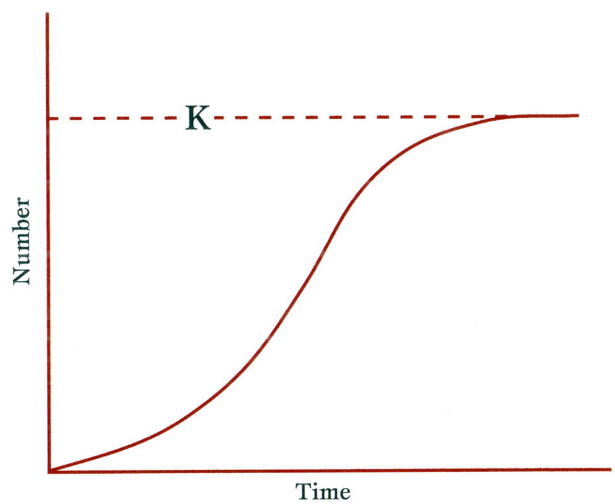

FIGURE 18.3. *S-shaped Population Growth Showing K* (carrying capacity).

C. Competition with members of the same species, known as **intraspecific competition**. The following are examples of intraspecific competition:

1. All the food is consumed and individuals begin to starve.
2. Individuals compete for what are known as "safety sites" (hiding places) and those that lose are likely to become prey or die due to adverse weather.
3. Toxic wastes begin to build up and poison individuals.

Density dependent factors frequently result in cycles of populations. Examples include rabbits and coyotes and in the insect world aphids and ladybird beetles. That is the reason insect control specialists resist spraying when predatory insect populations begin to build up because they understand that nature has a way of limiting population growth.

Populations also regulate themselves by limiting the number of new individuals. For instance, in one study using the cabbage aphid when the aphid density was high, the population produced an average of 0.41 offspring per female compared with a population that was low where 17.58 offspring were produced per female.

Average fertility (eggs laid/female) is a key to population dynamics of insects. Some insects with the average number of eggs per female are: House cricket (1,060), Bed bug (541), Lacewing (477), Armyworm (1,451), Fruit fly (1,532), Ladybird Beetle (814), Honey bee Queen (120,000/year), Termite Queens of Africa (31.5 million/year).

Population Dynamics

1. Unchecked insect populations have what theoretical growth rate?
2. What is the "carrying capacity" of the environment?
3. What are the factors that limit population growth of an organism as it approaches "K"?
4. Why are insects marked and released relative to sampling?
5. What is a heat unit and how might it be used in insect control?

UNIT 5

Interaction with Humans

CHAPTER 19

Insect Products and Services

How Sweet It Is!

Honey has been called the food of gods and kings. Of course, folks of more earthly and less noble birth have also been known to partake of the "fruit of the hive."

Bees have been depicted in Egyptian hieroglyphics dating back to 3500 BC. Indeed, for thousands of years, honey was the primary sweetener used by mankind.

Honey is wonderful. It does not spoil. Its acidity and sugar content keep harmful microorganisms from growing in it. It does not freeze. And, as people found out in times before refrigeration, when used on salt-cured foods, it covered up some of the taste of the salt.

Honey bees were unknown in North America until the immigration of Europeans. The introduction of the honey bee prompted the Native Americans to dub these insects "the white man's flies." Nonetheless, Native Americans quickly acquired a liking for honey.

Honey truly results, as an old proverb teaches, from the industry of the bee. It takes 2,000 to 3,000 bees working an entire day to collect enough nectar to make a pound of honey. During the day, each worker bee makes six to eight trips to the field to visit the flowers. She will suck the nectar from the flowers and store it in her crop, a specialized portion of her digestive tract, for the trip to the hive.

Once in the hive, she regurgitates the nectar, which has undergone a chemical transformation while in her crop. The honey is then placed in a cell of the honeycomb, where it is fanned until the water content reaches a proper level for storage.

Honey varies widely in color and flavor, reflecting the type of flowers that provide the nectar. In general, clover plants provide a very sweet and light-colored honey. Honey from the roadside weed goldenrod is very dark in color and not as sweet. Some liken goldenrod honey to sorghum.

Each honey, like a fine wine, has its own bouquet. From orange honey can be detected the odor of, well, orange blossoms. Not surprising because it's the nectar that gives the characteristic aroma to blossoms of all types.

Honey has also been used to make wine, called mead, which was a favorite drink of the Knights of Merry Olde England. In the days of the Middle Ages, it was also customary to provide a newly married couple with enough mead, or honey wine, to last for the first month of their married life. Since a month was roughly equivalent to a cycle of the moon, this time became known as the honeymoon.

So it is entirely appropriate that the most famous line from the long-running TV show *The Honeymooners* should be Jackie Gleason's "How Sweet it Is!"

Flowers Market Themselves to Insects

April showers bring May flowers. Or so the old saying goes. May flowers are not the first flowers of the season. However, it is during the month of May that flowering plants go into full swing for the growing season.

We humans appreciate flowers. So we look forward to the time each year when the beautiful blossoms of the plant world adorn the landscape.

In our adoration of the beauty of flowers, we are tempted to believe that Mother Nature placed them there for us. But think again! The flowers are there for insects. We humans are merely spectators to a great marketing scheme in nature—marketing for the sake of pollination.

Pollination is the process whereby plants reproduce themselves. Pollination of plants results in the production of seeds. Seeds germinate and grow into plants. And plants produce more seeds. And so it goes in one of the endless cycles so common in nature.

Plant pollination sometimes depends on outside forces to move the pollen from place to place. The wind does the job for some plants. Other plants depend on insects to transport the pollen from flower to flower.

Insects, like most things in nature, can be described as selfish. They don't work for nothing. Plants pay the insects to carry pollen by offering a little sip of nectar—a sweet bribe to labor for the plant's benefit.

It is not enough for plants just to provide nectar. In nature, as in the modern business world, sales and marketing are driving forces. So plants have evolved elaborate schemes—flowers—to entice insect pollen carriers.

The bright colors of flowers are billboards and neon signs for insects. To a flying insect, a meadow in flower must appear much like the Las Vegas strip to incoming airline passengers. Each flower radiates an enticing message: Stop here!

But there is more. Frequently, the flower emits a wonderful aroma—an aroma that further entices the insect visitor to stop for a sip. An aroma that to an insect is as enticing as a hawker at a Las Vegas casino. But as Friedrich Barth says in his book, *Insects and Flowers*, "it is not enough to just get the insect to stop at the establishment, one must still find the door."

So like any good business, the flower has signs to the entrance. These signs are known scientifically as nectar guides. Nectar guides are frequently associated with ultraviolet reflections on the flower. These are unseen by humans but are blinking neon signs to insects that see well in the ultraviolet portion of the light spectrum. The nectar guide is saying to the insect: follow this path to the sweets hidden deep in the flower.

The path to the nectar results in the insect contacting pollen. Some of the pollen is trapped on the insect's body. For the flower, it has accomplished the first marketing goal. The insect is now a pollen carrier.

There is more to the story. For pollination to be completed, the insect must carry the pollen to another flower. For this reason, flowers just give out small doses of nectar. The insect, frequently a bee, must go to other flowers to get all the nectar it can hold. When this happens, the pollen from one flower is transported to another. To flowers, it is now mission accomplished!

So the next time you stop to smell the roses, remember that you are benefiting from a close association between insects and flowers. It's an elaborate marketing scheme perfected by flowers involving advertising, supply control and repeat customers!

Forensic Entomology

An entomological who done it!

We all know that insects can cause problems. They eat our plants, animals, possessions and even feed directly upon us. Sometimes when we are bitten, or our animals, plants or possessions have been damaged, it is not clear what kind of a creature caused the problem. If the perpetrator of the crime remains at the scene long enough to be caught—for example when we swat a mosquito or dig up a grub in our lawn—it is easy to determine the culprit. However in many situations, by the time we notice damage the insect is long gone. The question is a common one, "Who done it?"

Solving mysteries has always played a prominent role in law enforcement. Most criminals are long gone when a crime is discovered. So crime scene investigations involve the collection of evidence that will aid in determining who committed the crime.

Society in general is fascinated with mysteries and criminal investigations. Our bookstores are filled with mysteries including Sherlock Holmes, Nancy Drew and the Hardy Boys. By the millions we have watched on TV—Magnum PI, Hawaii Five-0, Jake and the Fat Man, Mike Hammer, Cagney &

Fruit Flies Have Great Genes

What has 10 wheels and flies? Before you start calculating the number of wheels on your favorite airplane, think insects. The answer to that old grammar school joke is: a garbage truck. Most of us recognize the implication of the joke that, at least in warm weather, garbage and flies are inseparable.

Some of those flies hanging around the garbage might just be a group of flies known as *Drosophila,* the fruit flies. These little flies, only 3 to 4 millimeters in length, are named for the fact that they live on decaying fruit. There are many kinds of fruit flies, and they are very common insects. So common that it is nearly impossible to have ripe fruit lying around for any length of time without attracting a few of the little creatures.

While fruit flies are a nuisance to most of us, biologists have made good use of one species, *Drosophila melanagaster.* For many years, fruit flies have been the standard animal used in genetics research and teaching. The flies are easy to grow, reproduce rapidly and have big chromosomes.

Scientists over the years have developed many strains of this fruit fly, all which have different physical characteristics. Some of these strains can be purchased through biological supply houses for use in teaching genetics in biology.

Several eye colors exist. Regular old fruit flies, those that show up around the overripe banana on your kitchen counter, have red eyes. You can get flies with clear ruby eyes, or vermilion or white or apricot or brown. You can even get a strain of fly that has red eyes that dull with age.

Normally the fruit fly holds its wings along its back when it is at rest. But scientists have produced a strain that holds its wings at right angles to the body. You can get a fly with almost no wings, called vestigial winged, or one with no wings all. Some have wings that curl upward. On another strain the wings are truncated.

If you don't like the normal yellowish color, you can buy a fruit fly that is black. Or you can get one with forked bristles on its body; normal ones have straight bristles.

All of this makes for some interesting combinations and descriptions of fruit flies. For instance, you can buy a strain of fly that has white eyes, miniature wings and forked bristles. Or you can get one that has curly wings and plum eyes with divergent wings and short-and-thick bristles. Now I'm not sure what you get if you cross a forked-bristle fly with a short-and-thick-bristle fly. It would make a good experiment in a science classroom, and that is how these insects are used.

For most of us, the only fruit flies we are interested in are those that are sold as the wild type—described as normal characteristics with a red eye color. Those are the ones we find in our kitchens. In general, the description we most appreciate is one that goes like this: small blob of yellow insect smashed on countertop!

Lacey, Simon & Simon, Kojak, Columbo, Hart to Hart, Knight Rider, Matlock, Get Smart, CSI, Law and Order, Diagnosis Murder, and Murder She Wrote to name a few.

Another of the famous TV crime solvers was Quincy. Quincy was officially billed as a **forensic** pathologist. Quincy gathered all kinds of evidence from murder scenes and through his laboratory tried to determine the facts relative to the case. The word forensic comes from the Latin *forensis* from forum or public place. By definition forensic means "belonging to courts of judicature or to public discussion and debate." Thus debate and speech teams are sometimes known as forensics clubs. Relative to crime investigations forensic means that the evidence will be presented in public, as in a court of law.

In recent years the concept of forensic entomology has gained prominence. Forensic entomologists use insect evidence to determine facts surrounding a crime scene. While using insects to help solve a murder is an appealing one to the press, the idea of using facts to pinpoint the cause of a problem that might have been due to insects has long been an important aspect of entomology.

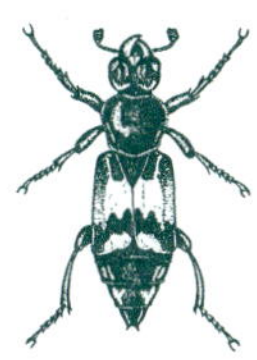

In "whodunits" as in police work all the evidence collected must stand up in a court of law, since that is always the potential final outcome of a case. This in general means collecting and logging samples and taking photos. An important aspect of evidence photos is that they must be able to be identified as to location. This means landmarks in the background that can be used to place the photo. It is also good planning to include involved parties (landowners, farmers, chemical dealers) in some photos.

For police and entomologists the easiest way to solve a crime is to catch the criminal (insect pest or human) at the scene. However, investigators must make sure that the suspect is the actual criminal. This is especially true of insects in a field that has been damaged by something—almost any insect that happens to be there is an immediate suspect!

If you can't catch the criminal the next best way is to be able to place the suspect at the scene. This can be done by personal belongings. For insects personal belongings could be things like pupal cases or frass. Indeed insect frass is sometimes a good indicator of just what insect, if any, has been at the scene. Another good indicator used to detect human criminals would be fingerprints or blood. For the insect a similar indicator would be a feeding pattern, the type of chewing marks or whether the feeding was on leaves or stems.

Good investigators of insect damage also like to consider the pattern of damage in fields being considered. For instance, insect damage can sometimes be confused with equipment malfunction (such as planting problems), soil type, or water drainage patterns. One thing to remember in such situations is that insects seldom follow down single rows, something that can be related to equipment functioning.

A far more visible aspect of forensic entomology is the use of insects to help solve crimes related to humans. This is most commonly used to solve murders. Most frequently, insects are used to determine the time of death. In general, time of death in forensic investigations is determined through one of four general methods:

1. **Histological methods**—based on structure changes of tissue.
2. **Chemical methods**—based on chemical changes in the body.
3. **Bacteriological methods**—based on microbial changes.
4. **Zoological methods**—based on plants and animals including insects.

Research using pigs as the model animal has shown that there is a succession of insects that invade an animal carcass following death. These come in what is known as waves. The waves occur over a period of years. The first wave includes the blow flies. These flies show up almost as soon as death occurs. It has been said that a blow fly is the first creature to show up at a crime scene other than the perpetrator. In the middle waves, once the fluids are gone from a body, the sexton beetles show up. In the last waves when the only things that remain are hair and bones, the insects that occur are the dermestid beetles and the mealworms (*Tenebrio*). It may take as long as 3 years to reach this stage.

Using such information, entomologists can place the time of death by determining the development of blow fly maggots. By knowing the species of blow fly and the temperature, it is possible to calculate precisely how long the insect has been developing. Thus the time of death can be determined. Consequently, police investigators are being trained to take notice of insects around a crime scene—they may have an interesting story to tell!

Forensic Entomology

1. What are the first insects to arrive on a carcass?
2. What animal is used as a model for human decomposition and insect infestation?
3. What is forensic entomology?
4. What meteorological factor must be considered in addition to insect development in estimating postmortem interval?

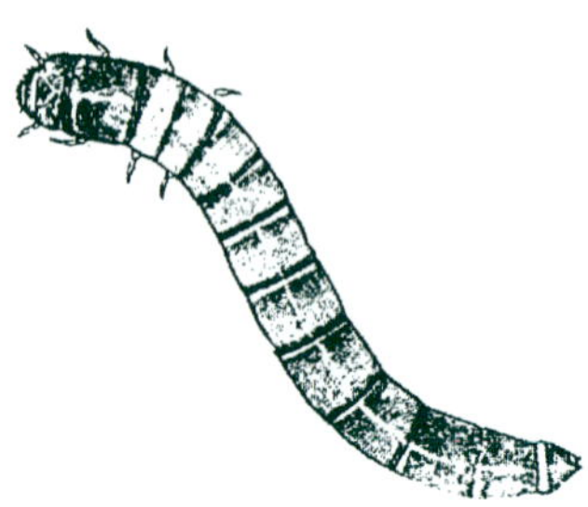

CHAPTER 20

Insect Pests and Control

Ecologically speaking there is no such thing as pests, only consumers. However, when some organism begins to consume something that mankind wants, that organism is branded a pest. From the beginning of agriculture, pests have created problems for farmers. Early farmers learned to live with pests and, through trial and error, how to deal with the more troublesome ones. Ancient Chinese citrus growers maintained colonies of a predatory ant to place in orange trees to reduce the numbers of leaf-feeding insects. Around 1,000 BC, Homer mentioned "pest averting sulfur" as a method of insect control. Long before the discovery of America, Indians in Central America hilled corn to prevent lodging due to the feeding of corn rootworm larvae. Many of the ancient control practices used techniques such as crop destruction, tillage and crop rotation.

In the latter part of the 1800s Paris green and rotenone were discovered to have insecticidal properties. In 1939, Paul Muller established the insecticidal value of DDT and, in 1948, won a Nobel Prize for his efforts. The use of DDT as an insecticide and the development of dinitro compounds such as DNOC and dimoseb as herbicides signaled the beginning of the chemical age of pest control. From 1950 until 1975, total production of synthetic organic pesticides in the U.S. increased at a steady rate. Pest control literature during this time deals mainly with chemicals. This period was truly the "golden age" of chemical pest control.

In 1962, Rachel Carson published *Silent Spring*, a book that dramatized, some say a bit too dramatically, the perils of excessive pesticide use. *Silent Spring* served as a focal point for the general public to push for a change in philosophy regarding the nature of pest control. In 1963, the President's Science Advisory Committee issued a special report that criticized a number of plant protection chemicals, primarily insecticides. Problems regarding chemicals began to surface. Target pests were becoming resistant. New pest organisms that had not existed previously appeared. Chemicals were found to be polluting streams and lakes and were detected in food. The golden age of pesticides began to appear less golden.

In 1972, the president of the United States directed agricultural, public health, and environmental protection agencies of the federal government to take immediate action to develop pest management programs. The goals of the programs were to protect the nation's food supply against the ravages of pests, to protect the health of the population, and to protect the environment.

This federal action signaled the beginning of the pest management era of pest control. This era has been accompanied by intensive regulation of pesticide manufacture and use by various federal agencies including FDA and EPA. Federal money was made available to fund pest management research. The Cooperative Extension Service provided money to establish pest management action programs in all states. The action programs were designed to demonstrate the value of pest management philosophy and associated control tactics for pests of all types. In addition, the federal program which required certification of applicators of certain "restricted-use" pesticides was an attempt to train users about the hazards of pesticide use.

The *Bug Scout* Cartoon— A Teaching Approach

In the late 1960s and early 1970s the concept of controlling insects with a minimum use of insecticides was an ongoing debate among insect control specialists, agriculturalists and environmentalists. The concept now known as pest management was being talked about by administrators at all levels but not really being used at the producer level.

Our Immigrant Insects

Have you ever wondered why many pest insects have foreign names?

It sometimes appears that even insects contribute to the U.S. trade deficit. However, foreign insects began invading our shores long before Washington bureaucrats fretted about the balance of trade. Unfortunately, many of these uninvited insects have stayed here.

German and Oriental cockroaches have infested our homes for many years. Now, in some parts of the United States, the Asian roach has moved in. Many gardeners find Japanese beetles munching on their roses during the summer. Wheat farmers have to deal with Hessian flies and Russian aphids. Corn is attacked by the European corn borer, and soybeans, by the Mexican bean beetle. Apple producers battle the European red mite and the Oriental fruit moth.

American elm trees are now rare because of a disease organism carried by the European bark beetle. All of us have read about the Oriental and Mediterranean fruit flies. Now the African bee has entered the U.S. from Central America.

All of these insects are named for their native areas. When mankind began to travel worldwide, insects hitched a ride. The Hessian fly probably came to the United States in straw bedding used by the Hessian troops during the Revolutionary War. Japanese beetles willingly ride airplanes—many new infestations begin near airports. The Mediterranean fruit fly sneaks into the country concealed in fruit, which is sometimes concealed by travelers.

Some insects that cause problems were actually "imported." For instance, the African bee was introduced to Brazil in an attempt to increase honey production. Gypsy moths, pests of trees, were introduced into the United States by a scientist working on silk production.

All introduced insects are not bad. Some beneficial insects have been successfully introduced into the United States. One such insect is the Vedalia ladybird beetle—a predator of a citrus insect pest.

A few insects native to the United States have managed to find their way to other lands. The American bollworm and the Colorado potato beetle are now worrying farmers outside the U.S. When it comes to pest insects moving around the world, some might say, "share and share alike!"

Agricultural educators faced a difficult problem: how to get growers to think about the concept of pest management. The problem was difficult because insecticide use was supported by a powerful pesticide industry that used modern marketing and advertising techniques. Pest management on the other hand was supported primarily by the public sector, which did not have access to the mass media in the same way that the pesticide industry did.

Out of this dilemma grew new ways to reach the growers with the idea of pest management. One such idea was the development of the cartoon character now known as "Bug Scout" (Figure 20.1). Bug Scout was based on the stereotypical image of an entomologist on a field trip—short pants, pith helmet and insect net. Scouting was the idea that it was necessary to "scout" crops—go through the field looking for insects and their damage. The character was combined with a short Pest Management Tip column and introduced into a farm magazine *Indiana Prairie Farmer* on a regular basis.

The rationale for the cartoon and short text format was that most readers of magazines will look at cartoons and will only read a short article unless the material is of a great deal of interest to them. It has also been suggested that the same is true of college students!

The experience with Bug Scout and Pest Management Tip proved it to be a good teaching tool. For instance, in the first readership survey of the magazine that included the feature, it was one of the highest readership items. About 62% of the readers read "Bug Scout" compared to an average

FIGURE 20.1. Bug Scout cartoon used to promote pest management.

Ladybug, Ladybug Get Out of My Home

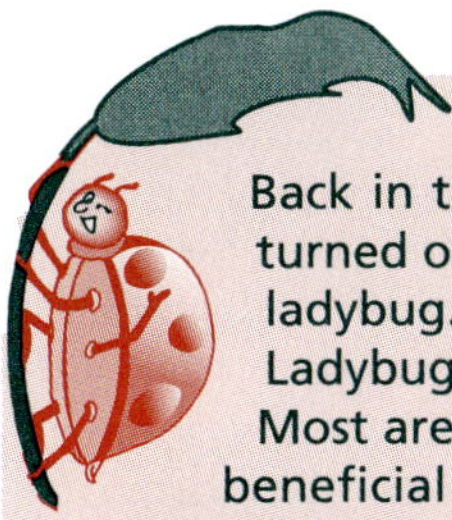

Back in the 1920s, a cockroach poet named archy wrote a poem called "the froward ladybug." As it turned out, the ladybug that had become a problem for archy was not the insect most of us know as a ladybug. The ladybug of archy was a female insect, a lady bug.

Ladybugs are some of the most recognizable insects. They are beetles with hard shells or wing covers. Most are colored bright orange or red. Many have dots on their wing covers, and most are considered beneficial insects because they are well-known predators. They feed on aphids, plant-eating insects hated by most farmers and gardeners. And in spite of their name not all ladybugs are ladies. Many are gentleman bugs!

But even an insect with the name of ladybug that gets rid of insects we don't like can sometimes wear out its welcome. In the case of ladybugs, that occurs when they move into our homes during the winter months. When they do, most home owners take a dismal view of the activity, and like the cockroach archy, they acquire a dislike for the ladybug.

During the winter, ladybugs in the adult or beetle stage seek protected sites in which to hibernate. Such places include under rocks, piles of leaves, in window wells and under the eaves of houses. Most ladybugs, however, appear not to be content to stay in protected places outside. Instead, they manage to crawl into our attics or wall spaces. From there they find their way into the interior of our warm homes and crawl over cabinets, carpets and light fixtures.

Wintertime ladybug invaders vary in color, size and in number of black spots. These differences lead many folks to conclude that several ladybug species are in their homes. That may or may not be true.

There are several hundred species of ladybugs, many of which are found in homes at one time or another. However, home-invading ladybugs usually are the species known as the Asian ladybug. Asian ladybugs vary widely in color and markings even though they are of the same species.

Ladybugs do not cause damage in our homes so their presence is a nuisance rather than a real problem. They will not eat our houseplants or damage our clothing. They do not often bite, although they might if held tightly.

They do have a rather nasty characteristic however. They sometimes smell bad. Not all the time; just sometimes. The stink comes from the scientific habit known as reflex bleeding. When the insect is disturbed, it can expel a smelly, yellowish liquid from the area where the leg joins the body. The insect uses the liquid to ward off predators because it tastes as bad as it smells. So, when we touch a ladybug it sometimes responds as if we were a predator and squirts us with the ladybug equivalent of mace.

When that happens, we, like archy the cockroach, could describe the insect as froward, or—as Webster defines the word—obstinately willful!

of 49% for other cartoons. The pest management tip was read by 46%. This compared to an average readership of 30% for special and regular features of the magazine. Advertisements also attracted about a 30% readership. In general, Bug Scout and the Pest Management Tip were a good way to reach growers with pest management information. Furthermore, they were quite cost effective—the magazine actually paid for the cartoon and published the information free.

Insects and Germs

A Lousy Subject

Few humans could love a louse, especially one of the insect variety.

Most of us find these little creatures quite distasteful. So much so that when we feel a bit under the weather we are apt to describe the condition as "lousy." Many of us have really "loused something up" at some time or another, and all of us have heard a person described as "a real louse."

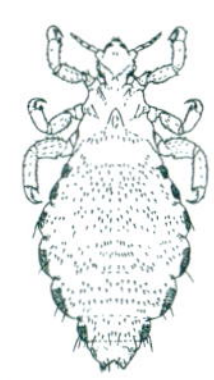

What are these insects that have come to symbolize the despicable aspects in our lives? Lice can be described as wingless parasites of warm-blooded animals. Hosts include humans and that is where the rub comes in! To have lice is something to be avoided, or at least not mentioned publicly.

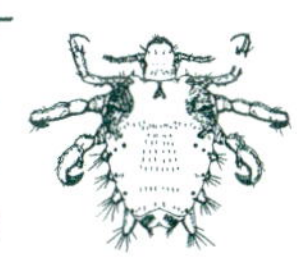

It has not always been so. You see humans have harbored lice for as long as, well, as long as we have been humans. The oldest literature makes reference to lice, including several references in the Bible. Mummies of Native Americans, estimated to be several thousand years old, show evidence of lice infestations.

The Domestic Fly Is No Saint

The house fly *Musca domestica* has been an associate of humankind throughout recorded history. From the times of the Bible, "And there came a grievous swarm of flies into the house of Pharaoh . . ." (Exodus 8:24), and Aristotle, who is said to have been fascinated by the mating behavior of these two-winged insects, to the 20th century, almost every person recognizes a fly when they see one.

Even the species name, *domestica*, suggests that the house fly lives in close contact with humans. Some authorities believe that the association of house flies with people is more than a superficial relationship. These insects depend on us for subsistence.

Immature house flies are ecologically known as saprophytes, living organisms that feed on dead and decaying things. House fly food includes garbage and animal and human manure. Historically, house fly populations are high when there is an abundance of food for their larvae. Consequently, as sanitation practices improve, house flies are less abundant.

In the United States, house fly populations decreased in many areas as sewage systems developed and replaced outdoor privies. In addition, transportation changed from the horse to the horseless carriage, and a fly-breeding site—horse manure at the side of the road—became scarce. So there are probably fewer house flies today than when Grandma and Grandpa were kids.

Having fewer house flies is something most of us can tolerate. Indeed, the house fly can carry disease organisms on each of its six feet. For that reason, an effort was made at the turn of the century to establish the name "typhoid fly" for what we know as the house fly. The reason was that typhoid fever was one of the most serious diseases carried by this insect. Luther West, in his book *The Housefly*, suggests that "cholera fly" or "dysentery fly" also would be appropriate names. These names recognize that the insect can transmit a number of diseases as it moves from breeding sites to our dinner table, where it happily walks around on our food.

Window screens and fly swatters came into widespread use during the early part of the 1900s, partially in an effort to keep house flies from contaminating our food. Fly swatters were popular advertising items during this time. Many swatters included such slogans as "swat the fly and prevent disease."

While modern sanitation and insecticides have reduced the potential of house-fly borne diseases in modern societies, the problem still exists in many developing areas. But even today we are sometimes reminded of the past when "shoofly pies" were baked, a pie so named because it was necessary to shoo the flies from the pies as they cooled on the windowsill—a good idea because the flies had no doubt come from the privy or the barnyard. And the thought of having them tromping in our pies was not very appetizing!

During the Middle Ages, an infestation of lice was considered a sign of good health, an odd notion since lice were responsible for transmitting the disease typhus. Folks probably developed this idea because lice and fleas leave a body very quickly at death—sort of the insect equivalent of rats off a sinking ship.

So prevalent was the association of lice with health that servants, as a sign of hospitality, would capture and toss lice on guests as they passed through the gates of the castle to attend a royal gala. Ladies of fashion of such times used long hair pins to keep their tresses in place. A practical use of these fashion items, it is said, was to dislodge particularly troublesome lice should the creatures manage to create discomfort near the scalp.

Lice also played a political role in the olden days. In Hurdenburg, Sweden, during the Middle Ages, a louse was used to determine the next mayor. Each year the eligible notable men of the town would sit at a round table with their beards touching it. A louse dropped in the center would determine the next mayor by selecting a beard for a home. Sort of the ancient electoral college, one is to presume.

Lice are still with us. Each year outbreaks of head lice are experienced in schools around the country. These outbreaks, which are most common during winter months, are frequently found among school children in the lower grades. The reason for this is probably the tendency of young children to share winter clothing, especially stocking caps. This allows lice to travel from person to person. Additionally, these youngsters are at an age when parents encourage children to groom themselves, a task which is accomplished a bit haphazardly at times.

Indeed, the great apes and ancient humans engaged in mutual grooming, the primary purpose of which was to remove lice eggs from the hair. These eggs are called "nits," and the practice of hunting them is called "nit-picking."

So remember the next time you accuse someone of being a "nit-picker," it's probably because they have a lousy job!

Flea-Bitten

Have you heard the old saying, "Dogs have fleas to keep them from worrying about being dogs?" Well, it's true that dogs spend a lot of time scratching, and for good reason: flea bites itch. Fleas can cause their host animals to be unthrifty. Thus, another saying, "The fatter the fleas, the leaner the dog."

Fleas are among the most modern of insects. Some scientists believe that fleas are just flies that have lost their wings. Fleas live on mammals. So they can move easily through the hairs on their hosts, fleas are flattened from side to side.

Adult fleas are known for their jumping ability. A flea can leap up to 13 inches. If a human could do as well, the world long-jump record would be nearly 800 feet. But such a comparison is not fair. Fleas have rubber-like pads at the base of their back legs. These pads are compressed when the flea is preparing to leap. Thus, flea jumps are spring-aided!

Immature fleas live in the nest of the host and feed on feces and other organic materials. Before changing into adults, young fleas enter a resting stage that can last for months. The fleas emerge from the resting stage when a host animal enters the nest. Either the motion or the heat from the host's body triggers the emergence. Such behavior ensures that a food source is available to the newly-emerged and hungry flea.

Some fleas are named after specific animals such as the cat flea and the dog flea. However, most fleas will feed on any warm-blooded animal. In fact, the most common flea found on dogs is the cat flea, not the dog flea.

It is hard to be positive about fleas, but they have been trained to pull carts and perform in flea circuses. The training is the easy part, but finding a harness to fit a flea is another matter.

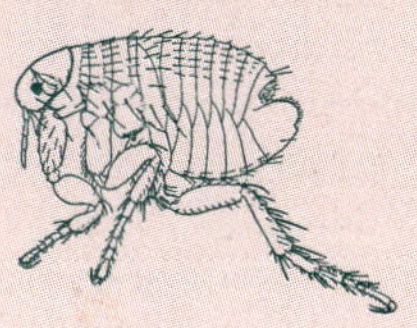

Humans have long been bothered by fleas. In the Middle Ages, ladies even wore "flea furs." This piece of fur was worn around the neck as an attraction for fleas. The collar with its collection of fleas could then be removed and the fleas destroyed.

Ogden Nash recognized the long association of humans and fleas in his poem entitled "Fleas." The entire poem is "Adam Had'em."

A Billion-Dollar Insect

This is a tale of a big tree, a little insect, and a fungus you can't even see. The big tree is the American elm. Once upon a time, mature American elms were common in the United States from the Missouri River to the Eastern Seaboard. Majestic elms lined the streets of cities and towns, sometimes forming living tunnels with their arching branches. But no more.

Where have the elms gone? Most have been killed by a fungal disease, *Ceratocystis ulmi.* Early symptoms of infected trees are wilting leaves on some branches. These leaves turn brown prematurely and drop from the tree. Larger limbs produce numerous trunk suckers. Finally, in two to three years the tree dies.

The bark of dead elms killed by the fungus soon sloughs off, leaving the white wood of the tree exposed. These skeleton trees stand for many years in woodlots and on creek banks in stark testimony to days gone by, when American elms dominated the treescape.

An investigation of the dead wood or sloughed bark of dead trees reveals a series of tunnels in the sapwood. Each group consists of a main tunnel with a number of secondary tunnels running at right angles. These are caused by the feeding activity of an insect, one of the bark beetles that get their name from the habit of feeding just under the bark of the tree.

The particular bark beetle associated with the death of the elm trees is known as the smaller European elm bark beetle. Feeding by the bark beetle will not directly cause the death of the trees, but as the insect moves from tree to tree it transports the fungus that does kill them.

Ceratocystis ulmi also is known as Dutch elm disease. And for good reason. It was first identified in the Netherlands in 1921. Since then, it has spread throughout Europe and is now found in parts of Asia and North America. Dutch elm disease was first identified in the United States in 1930. Both the fungus and the smaller European elm bark beetle came to the states in elm logs to be used for furniture veneer.

The beetles overwinter in the tunnels, where they spend their life as larvae, and emerge through small round holes in the bark in spring. At that time, they fly to the growing points of the newly leafed trees to feed and, as a result, transmit the fungus to another tree. Beetles do not fly great distances—usually not over 200 yards—but they have been known to hitchhike great distances in vehicles.

While the beetles are the primary way that trees come down with the disease, it is not the only way. Trees that grow close together also can transmit the fungus when roots grow into each other. That is why large trees growing within 20 feet of a diseased tree also become diseased even if beetles aren't present.

Just think what a little beetle can do, at least when it transmits a disease organism. In this case, losses have exceeded $1 billion in the United States!

Insect Products/Insect Pests and Their Control/ Insects and Germs

1. Why do harmful organisms not grow in honey?
2. What alcoholic beverage is made from honey?
3. Where did the term honeymoon originate?
4. What does the foreign name of an insect generally indicate?
5. Why were gypsy moths introduced into the United States?
6. For what general purpose are insects deliberately introduced into other countries?
7. Typhus is transmitted by what insect?
8. What are lice eggs called?

Insecticides

Chemical Safety Starts with Mice and Men

Don't underestimate the value of rats and mice! As farm pests they are quite undesirable. As laboratory animals they can be used to estimate the toxicity of agricultural chemicals to humans. This information is necessary to determine safety precautions needed when handling toxic chemicals.

The benchmark for measuring the toxicity of a compound is LD50. An LD50 is the dosage required to kill 50% of the test animals. Lethal doses can be based on oral or dermal exposure, depending on how the chemical was administered to the animals.

Lethal dose information is used to determine the signal words required on pesticide labels. These signal words—caution, warning, or danger—indicate the toxicity of the pesticide.

The next time you look at the toxicity of a pesticide, remember that mice and men might have more in common than you thought!

BUG SCOUT

"How was I to know that this rodent was the only survivor of your insecticide toxicity test?"

Pesticides are Poisons

All pesticides are poisons. However, pesticides are not equally hazardous to humans. Thus, federal law requires that all pesticide labels include a signal word. The signal word indicates the poison category for the product.

The most hazardous compounds have the signal word "Danger" and a skull and crossbones on the label. Moderately hazardous pesticides have the signal word "Warning." The least hazardous of the pesticides have the signal word "Caution."

These signal words are meant as a guide to the user. However, regardless of the signal word, all pesticides should be handled carefully. People who handle pesticides should know the location of the nearest poison control center. That knowledge might be valuable in case of pesticide poisoning.

Remember, you can never be too careful when handling pesticides. Treat pesticides as if they are poison. They are!

BUG SCOUT

"The skull and crossbones? Oh, it's used to signal danger. What I mean is . . . Well, we're talking about insects!"

Pesticide Cover-Up Good for Health

Farmers handle more pesticides during the spring and summer months than any other time during the year. More handling of pesticides means more exposure and more potential for accidents.

You can minimize pesticide accidents by following these suggestions:

- Do not smoke or eat while applying pesticides.
- Mix pesticides carefully to avoid splashing.
- Avoid breaks in pesticide containers.
- Wash with soap and water after using pesticides and launder clothes before wearing again.
- Wear protective equipment as prescribed on the label.

Remember that the use of common sense around pesticides is good advice. Many government officials will tell you that suggesting a "cover-up" is bad advice. However, when handling pesticides the smart farmer will "cover up" every chance he gets!

BUG SCOUT

"Safety with pesticides is important, but don't you think the suit of armor is going a bit far?"

Store Pesticides Safely

Storage of pesticides in unlabeled containers is a dangerous practice. Pesticides stored in feed bags frequently end up in livestock rations, often with disastrous results. A significant number of human poisonings each year also result from pesticides being stored in anything other than original containers.

Be safe! When cleaning pesticide application equipment, put any unused product back in original bags. In addition, always store pesticides away from your livestock feed. Insist that your feed dealer do the same with his pesticides.

A pesticide out of place is an accident just waiting to happen!

BUG SCOUT

"Let's see, I put the rootworm insecticide in the cattle feed bag . . . or was it the potato sack? Now the herbicide . . ."

Control Spray Drifts

Spray drift to non-target areas can be a problem when applying pesticides. Drift can be due either to vapors moving from the target area or due to movement of spray during application. Spray movement, however, is the primary cause of drift problems.

The amount of spray drift is due to droplet size, wind speed, and spray height above the ground or crop.

Drift can be minimized by eliminating the fine spray particles. This can be done be decreasing pressure or increasing water volume. However, more spray drift is due to wind problems than to any other factor. You can't do anything about the wind, but spraying in high wind is sure to create problems.

Remember, good fences and lack of spray drift makes good neighbors.

BUG SCOUT

"Oh, he's not spraying here . . . he's spraying way over there!"

Insecticides

1. What is an LD50?
2. Name the signal words that can be used on a pesticide label.
3. What are some suggestions useful to minimize pesticide accidents?
4. In what kind of container should pesticides be stored?
5. In addition to wind speed what are two other factors that will contribute to spray drift from a field application of pesticides?

Alternatives to Chemical Control

Pheromones Aid Insect Control

Pheromones are complex chemicals produced by insects. These chemicals are used by insects to communicate with other members of the same species. For instance, some female insects produce a pheromone that attracts males at mating time. Some pheromones are known to attract males over long distances—10 miles or more in some cases.

Scientists have been able to identify the chemical structures of many insect pheromones. Synthetic pheromones are now available for European cornborers and black cutworm moths.

Entomologists use synthetic pheromones to attract insects to traps. Insect counts from these traps are used to determine peak periods of insect activity. This information is helpful in timing scouting and control procedures.

Some entomologists thinks it's a dirty trick to use a sex pheromone to attract a moth to a trap. But some farmers think it's an even dirtier trick for cutworms or cornborers to attack a corn crop.

"Either they're attracted to the pheromone in the vial or your deodorant has failed!"

Consider Biological Control

Caterpillars of cabbage butterflies can cause severe damage to cabbage, cauliflower, or broccoli in the home garden.

Many chemical insecticides can be used for control of these caterpillars. Control can also be achieved with *Bacillus thuringiensis*, a species of bacteria that attacks and kills larvae of moths or butterflies.

Bacillus thuringiensis, also known as BT, is available under the trade names Bistrol, Dipel, and Thuricide. The use of a bio-control agent like *Bacillus thuringiensis* has advantages for the home gardener. There is no harvest restriction for the treated plants. BT is not toxic to honey bees and is safe to handle.

If the prospect of a few uninvited cabbage worms on the dinner plate isn't appetizing to you, *Bacillus thuringiensis* may be helpful.

"I don't care if your contract is for corn and soybeans, you're going to help me control cabbage worms!"

Crop Rotation Aids Insect Control

An old adage recommended planting three seeds in every hill of corn. The reason—"one for the grub, one for the crow, one for to grow."

Today grub damage to row crops is very low. Grubs are the immature forms of May beetles. The beetles lay their eggs in the soil of sod pastures or grassy areas in other crops. Newly hatched grubs feed on plant roots for one to three years before emerging as adults.

Grubs become a problem when corn follows sod in the crop rotation. Grubs then feed on corn plants and reduce the stand. Today, because very little corn is planted following sod, grub problems are very minor.

When it comes to grubs in corn, it's just not like it was in the good ol' days. Thank goodness!

BUG SCOUT

"It's one way to get rid of grubs!"

Clean Grain Bins

Clean and inspect grain bins before harvest. Don't forget combines, trucks, wagons, and other grain handling equipment. Stored grain insect infestations seldom originate in the field. Most insect infestations begin in old grain in and around the grain bins.

When cleaning reveals insect infestations, use a residual spray in grain bins and grain handling equipment. If an infestation is detected in old grain, fumigate with a recommended insecticide before adding new grain.

Don't forget to inspect grain monthly during storage. Remember when dealing with Indian meal moths, grain weevils, and other stored grain pests: "An ounce of prevention is worth a pound of insecticide."

BUG SCOUT

"Having Bug Scout inspect grain bins sure keeps the rats exercised!"

Good Sanitation Cuts Fly Problem

Sanitation is an important aspect of fly control. Flies found around farm buildings develop in moist manure or other wet decaying organic matter. Some flies can complete a generation, from egg to adult, in as little as 10 days during August. So, it is easy to see why fly populations appear almost overnight.

Insecticides can effectively kill flies. However, without proper sanitation, use of insecticides is a losing battle. Manure removal at least twice a week is needed to break the breeding cycle. Let wet manure dry and dispose of old damp straw or hay away from buildings.

You can probably control more flies with a pitchfork or manure shovel than with any insecticide on the market. And you really can't control flies without a spreader!

BUG SCOUT

"Hey, Bug Scout, from here it's hard to tell whether the flies are attracted to you or the manure pile!"

Reduce Overwintering Pests

Some pests survive unfavorable environmental periods in crop residue. For instance, the apple maggot and plum curculio cause infested fruit to drop from trees. The dropped fruit then serves as an overwintering site for these insect pests. Removing and destroying fruit is a good way to reduce populations of these pests.

The anthracnose fungus and the European cornborer both overwinter in cornstalks. Fall plowing reduces winter survival of these pests. Reduced tillage tends to increase infestations of pests that remain in crop stubble.

When it comes to managing pests, a good "housecleaning" can work wonders.

BUG SCOUT

"It could be caused by apple maggot or plum curculio or even gravity!"

Prune Fruit Trees Now

Fruit production may be affected by many pest problems. These pests include diseases, insects, mites, mice, and weeds. Proper pest identification is necessary to select an effective control program.

Pruning is an important aspect of most control programs. Removing and destroying dead or dying branches eliminates some sources of disease. Pruning also allows proper penetration of pesticides applied to the tree for disease, insect, or mite control.

Winter is a good time to prune fruit trees to aid pest control next summer. Remember a fruit tree is like a lot of organizations: a good pruning to cut out dead wood works wonders.

BUG SCOUT

"But you said to prune apple trees close!"

Alternatives to Chemical Control

1. For what insect management purpose are synthetic pheromones used?
2. Name a species of bacteria that is widely used for insect control.
3. Name three beneficial insects that are either predators or parasites of insect pests.
4. What is the most likely source of insects that start infestations in stored grain?
5. Other than insecticides, what is an important aspect of house fly control?
6. Why does removing and destroying fallen fruit in an orchard enhance insect control?
7. How does pruning trees affect the control achieved with pesticides?

Control in Home and Garden

Insect Devices Give Partial Control

There are a lot of gadgets on the market purported to provide insect control. These devices are sold to reduce insect problems in and around homes, barns, yards, or gardens. Most are designed to attract insects for purposes of extermination.

Other machines repel the undesirable insects. All such devices will provide satisfactory relief from insects only when used with other control measures.

No collection device will provide satisfactory fly control unless breeding sites are eliminated. The same is true for mosquitoes in the back yard. Thus, sanitation remains the number-one control measure for flies, even with the miracles of modern technology.

Remember, unless used with other controls, the only benefit of insect gadgets might be the pleasure of seeing and hearing flies commit suicide on the electric grid.

BUG SCOUT

"Not only is it approved by EPA, but it serves as a burglar alarm, starts your coffee maker, and . . ."

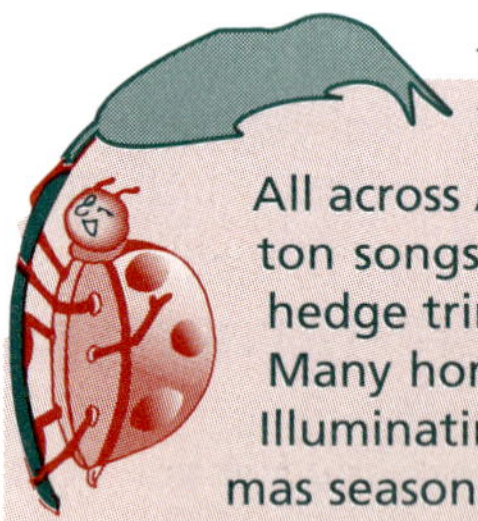

Bug Lights

All across America the signs of summer are evident. Wrens have returned to backyards to belt out two-ton songs from two-gram bodies. Neighborhoods drone with the weekend sounds of lawnmowers, hedge trimmers and sprinkler systems.

Many home owners also are engaged in another suburban ritual known as "hanging the bug light." Illuminating the bug light heralds the onset of summer as surely as lighting the tree signals the Christmas season.

In the good old summertime the age old battle between humans and insects reaches its peak. Bug lights (not to be confused with that popular beverage Bud Lite!) are another weapon in the war.

Although they vary in size and shape, all bug lights have a singular purpose: insect incineration! The device functions by using ultraviolet light to attract insects. Many night-flying insects are lured to light, but ultraviolet is the most attractive of all wavelengths. Modern technology has combined this light with open electric wires to form bug-killing devices known generally as bug zappers or bug killers.

Humans have flocked to these devices as surely as moths circle a flame because bug lights kill lots of insects. However, research has shown that the insects most home owners are trying to avoid are not affected by these electrifying devices.

Most female mosquitoes—the ones that bite and the ones we want to get rid of—are not attracted to the light.

In the case of Junebugs, many are killed but many more miss the trap. Subsequently, they land on vegetation in the area and lay eggs in the soil. These hatch into grubs that will feed in the lawn.

Generally, most of the insects that end up being killed in the light traps would not have been there if the light had not been hung in the first place. By using light traps we end up attracting insects, just the opposite of what home owners had hoped would happen.

So why do we pay good money for these devices? It's because we like to kill insects. There just is something pleasurable about the snap, crackle and pop of an insect being fried on the wires that gives us pleasure.

Nothing compares to sitting in a porch swing on a summer evening and listening to insects commit suicide in a bug light. But, hopefully, it is your neighbor's bug light. That way you know your insects are being attracted to the neighbor's place and not bugging you!

Pest Management

Introduction

Pest management is the current philosophy of agricultural pest control. As terminology, "pest management" is relatively new, having risen during the past decade to a place of prominence in the vocabulary of most agriculturists. Because it is a philosophy, pest management is difficult to define. Pest management is a concept of crop production incorporating effective and economical crop protection components that minimize the undesirable side effects of pest control actions. It is important to understand the following fundamental aspects of pest management.

1. Pest management is a part of total crop production.
2. Crop protection must be economical.
3. Pest control actions can result in undesirable effects.
4. In its broadest sense, pest management refers to all pests including weeds, plant pathogens, nematodes, vertebrates and insects.

Ecologically the term "pest" has no meaning. Mankind has classified certain organisms as pests because they compete for available food, destroy structures or carry disease. In agriculture, organisms were probably first recognized as "pests" when mankind began to cultivate crops. Under these conditions it was readily apparent that other organisms fed on crops and shared the harvest. As mankind progressed in agriculture and introduced monocultures, the impact of pests, under such unstable ecological systems, could be disastrous. In fact, many practices of modern agriculture may have evolved because of pest problems. Pests have probably influenced the course of civilization. The ancient Greeks and Romans knew of and used pesticides in the Mediterranean Basin. Crop losses due to pests during those times might have spelled the difference between adequate food supplies and famine.

During the last century agriculture in industrial countries has changed from small, labor-intensive to large, highly mechanized operations. As modern agricultural practices were widely adopted, pest problems frequently became more severe. In addition, the new mobility of the human population has allowed pests to move around the world and proliferate in areas free of natural controls.

As a result of the increasing need for pest control methods, several crop protection techniques have evolved. These include pest-resistant plants, cultural controls, biological controls and pesticides.

The success of the synthetic organic insecticides such as DDT following the conclusion of World War

Pest Management Isn't Poured Out of Cans

Almost everyone has a philosophy of pest control. Some people promote pesticides as the only approach to pest control. They have been described as being of the "kill 'em and count 'em" philosophy. Some people would never use a pesticide. Their philosophy has been termed that of an "eco-freak" or "bug and bunny lover."

A true pest management philosophy of pest control is somewhere between these extremes. The use of pesticides is an important tactic in pest management. Resistant plants and crop rotation as well as use of parasites and predators are also control options. Concern for the environment is important, as is economic benefit from control.

Remember, pest management is a philosophy. You can't go down to the local agri-chemical dealer and buy a can of pest management. When it comes to pest management, it's the philosophy that counts.

BUG SCOUT

"He said one-half pound in the row, with light incorporation, provides excellent control!"

II began the pesticide era of pest control. The new chemicals were effective, easy to use, and cost effective. As a result, most insect control programs quickly abandoned the technology that was currently being used in favor of pesticide spray programs. These spray programs led to the development of problems. Target insects began to develop strains which were genetically resistant to insecticides. Insecticides were destroying target insects and non-target organisms alike, and in cases where the non-target organisms were biological control agents, the pest population continued to increase in density. Pesticide residues began to build-up in ecosystems around the world, often ending up in the human food chain.

These and other events created an atmosphere of concern regarding the magnitude of use and dependence on chemical pesticides in pest control. These developments have led to a general recognition of the need for improved crop protection systems. The pest management concept and associated philosophy has developed in response to this need.

The Economics of Pest Management

Economic Injury Level

A basic concept underlying all pest management decisions is the economic injury level. Stern *et al.* (1959) defined the economic injury level as . . .

> *"the lowest pest population density that will cause economic damage."* Other definitions include a pest density where . . .
>
> *"the loss caused by the pest equals in value the cost of available control measures"* (National Academy of Science 1969);

"the pest population that produces incremental damage equal to the cost of preventing the damage" (Headley 1972). Regardless of how the economic injury level is defined, it boils down to whether or not treatment makes money for the producer, thus the term **ECONOMIC** injury level. For example, if the value of crop damage prevented by a pesticide application is $12 per acre and the cost associated with treatment (pesticide materials, application equipment costs and labor) is greater than $12, the pesticide application is not economical. We say that in this instance the pest population has not exceeded the economic injury level.

Find Economic Threshold

Pesticides don't increase crop yields. Pesticides can prevent yield losses when used against pest populations that reach damaging levels.

Pest managers refer to pest population levels where treatment is required as "economic thresholds." An economic threshold is the point at which the yield loss prevented will pay for the cost of treatment. Pesticide application to pest populations below the economic threshold means money wasted—straight out of the profit column.

Find out the economic thresholds for your crops. Thresholds vary according to the stage of plant growth, stage of the pest, and according to the type of growing season.

Some crop producers like to kill bugs or weeds every chance they get. Killing pests at populations below the economic threshold doesn't make "cents."

BUG SCOUT

*"Aha! One of the *!#!* little rascals is still kicking!"*

Economic Threshold

When the density of a pest population reaches the economic injury level, the crop has already sustained damage and loss in yield. In many cases this is a "tolerable loss" in that control would be uneconomical. However, when pest populations increase in density through time it becomes desirable to implement control procedures at a level below the economic injury level in those instances when pest populations will ultimately exceed the economic injury level.

The pest population level at which controls are employed to prevent the population from exceeding the economic injury level is the **ECONOMIC THRESHOLD** (Stern *et al.* 1959). In pest management the economic threshold is often referred to as the **ACTION THRESHOLD.**

Equilibrium Position

The economic injury level is always tied, in some way, to the pest population (Figure 20.2). Pest population density, like the population density of all organisms on earth, fluctuates over time. *The average population level of an organism over a period of time is termed the* **EQUILIBRIUM POSITION**. Changes in population numbers away from the equilibrium position can be caused by environmental factors (i.e., rainfall, adverse temperatures), biological control agents (i.e., parasites, disease) or competition between members of the population for food or space. Those populations that remain close to their equilibrium position are termed **STABLE POPULATIONS**, while those that fluctuate widely are termed **UNSTABLE**. Pest populations are frequently divided into groups according to the relationship of the equilibrium position and the economic injury level.

Economic Classes of Insects

Many insect species feed on crops but never reach densities high enough to cause economic injury. These insects are **NON-PESTS** and treatment is never required. Luckmann and Metcalf (1994) include the cow-pea aphid, *Aphis craccivora* Koch, on alfalfa; the yellow woolly bear, *Diacrisia virginica* (Fab.) on corn; and the painted lady, *Vanessa cardui* (L.), on soybeans as examples of non-pest insects.

Some insect species have very unstable populations and under favorable environmental conditions their numbers grow rapidly and exceed the economic injury level. These insects are termed **OCCASIONAL PESTS**. Many insects fit into this category of pests and include corn rootworm beetles, *Diabrotica* spp., feeding on corn silks; the green cloverworm, *Plathypena scabra* (Fab.) on soybeans or the fall webworm, *Hyphantria cunea* (Drury), on trees.

A number of insects have populations that nearly always exceed the economic injury level and are

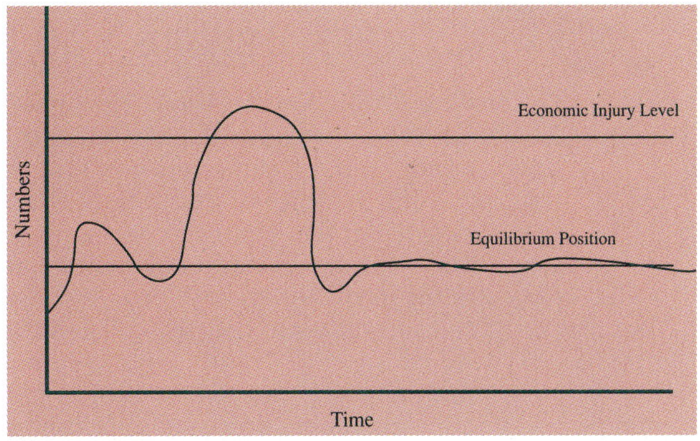

FIGURE 20.2. *Theoretical Concept of the Economic Injury Level Relative to Pest Numbers over Time.*

Base Control on Economic Injury Level

Pest management decisions are based on economic injury levels. The economic injury level is the number of pests needed to cause a loss in yield equal in value to the cost of control. Pest management specialists recommend control action when the economic injury level is reached.

The economic injury level is not constant. For instance, fewer pests are required to create an economic loss as the value of the crop increases. On the other hand, more pests would be required to reach the economic injury level as the cost of control increases. In addition, the number of pests required to cause economic damage changes as the plant grows and matures.

The economic injury level is just a way to indicate whether or not a pest control treatment will make money. Making money is the primary objective of crop producers and pest management tactics.

BUG SCOUT

Check Before You Spray

'Tis the good old summertime. That means bright sunshine, hot days and nights, and insects. Insects in our homes, our gardens, and our crops. The presence of insects does not automatically mean that control is necessary.

Some insects are beneficial. Lady beetles and lacewings destroy many aphids each day. Pest insect populations don't always reach damaging levels even though they feed on the crop. Control is not profitable under these conditions. Corn rootworm beetles feeding on silks of corn frequently do not require control.

Sometimes the damage is done before the insects are noticed. For instance, aphids on the tassels of corn have already damaged the plant. Aphid control at this stage is not cost effective.

When insects appear on your crops, decide whether control will be economical before spraying. Being quick on the spray can is important when control is needed, but don't be stampeded into spraying when it's unnecessary.

BUG SCOUT

"That masked hombre? The fastest spray can in the West, that's who!"

termed **PERENNIAL PESTS**. The cotton boll weevil, *Anthonomus grandis* Boheman; and the Colorado potato beetle, *Leptinotarsa decemlineata* (Say) are frequently used examples of perennial insect pests.

The final group of insects are termed **SEVERE PESTS**. Severe pests have an equilibrium position above the economic injury level and always require control intervention to prevent economic loss. The corn earworm, *Heliothis zea* (Boddie), on sweet corn; the codling moth, *Laspeyresia pomonella* (L.), on apples; and the house fly, *Musca domestica* (L.) in dairy barns are common examples.

Basis of the Economic Injury Level

The economic injury level is frequently based on a direct measure of the pest population; for example, insects per stem, or "grubs" per cow. However, many economic injury levels are based on indirect measures which are often easier to obtain than actual pest population counts. These indirect measures fall into two general categories:

1. The damage that is of interest is measured and includes a sampling for the percentage leaf defoliation, or the number of fruit with oviposition scars, etc.
2. The measure of a factor that is not in itself damaging to the plant, but which indicates the presence of a pest population that will create damage later. Leaf feeding by early instar European corn borer, *Ostrinia nubilalis* (Hübner), larvae on corn leaves is such an indicator.

It is important that a damage index, such as exists for the corn borer, be used in association with an estimate of the insect population. It is possible that unfavorable environmental conditions might reduce the pest population below damage thresholds. This would eliminate a control action, even though early leaf feeding exceeded the economic threshold indicated by the damage index. Indirect indices of pest damage can be fairly useful for determination of economic injury levels; however, their limitations should be considered when making control decisions.

Factors Influencing the Economic Injury Level

Economic injury levels are dynamic values. The value for any pest is influenced by a number of factors including value of commodity, cost of control, host plant tolerance, interaction of pest species and environmental stress on the crop. In general, as the cost of control measures and the tolerance of the crop to the pest increase, the economic injury level also increases. This means that as the cost of an insecticide treatment increases from $10 to $15 per acre, a producer would need a more damaging pest population before the yield savings would pay for the additional pesticide cost. Also, if tolerant varieties were used, larger pest populations would be required to justify treatment. Thus, the economic injury level goes up.

As the value of the commodity increases, the economic injury level goes down. Fewer soybean pests are needed to pay for the cost of treatment when soybeans sell for $8 per bushel than when they sell for $6 per bushel. In general, seed producers of any crop have lower economic injury levels than commercial growers. This is due primarily to the greater value of a unit of the crop. In addition, the plants used to produce seed might also be less tolerant than commercial varieties, thus reducing the economic injury level.

It is difficult to say that environmental stress and interaction of pests always affect the economic injury level in the same direction. As a general rule, increased stress on the plant, whether from the environment or other pests, tends to decrease the economic injury level. Treatment may be economically justified at lower pest levels when the plant is under moisture stress than when adequate moisture is available. The important point is that many factors influence the economic injury level and it may change from area to area, year to year, and even from field to field.

Zero-Damage Concept

A concept that is essential to a discussion of the economic injury level is what has been called the **ZERO-DAMAGE LEVEL**. *For every pest there exists a population level that will not have an impact on the yield or quality potential of the plant.* When the pest population is at or below this level, it no longer has pest status. Thus, considerations regarding changes in the economic injury level are no longer relevant. Unfortunately, many farmers have unnecessary production costs because they have become accustomed to killing pests, regardless of demonstrated need.

Furthermore, there is increasing evidence that populations at or below the zero-damage level may produce a positive yield response as shown with alfalfa and the alfalfa weevil in Indiana.

COUNT ROOTWORM BEETLES

Count beetles to help determine the need for rootworm control on next year's corn crop. Beetles present during August lay eggs that hatch into rootworm larvae next spring. An average of one beetle per plant is needed before the use of a soil insecticide is profitable.

Count the beetles on a minimum of 25 plants, five plants at each of five locations per field. Be sure to look behind leaf sheaths and in silks. For the best results, select plants that are typical of the crop and avoid sampling only in the border areas of your field.

Fields that do not have beetle populations and are in brown silk normally do not need further scouting. However, fields in green silk will attract beetles from neighboring fields and might require additional field visits.

A visit to a cornfield in August may not be a pleasure trip. But many corn farmers have found that it's a management tool that pays dividends.

BUG SCOUT

"Did ya count that one, Bug Scout?"

The Sporadic Nature of Certain Pests

The economic injury level is normally defined in terms of a plant or animal unit of production. However, it is equally important to consider the economic injury level in terms of the field or throughout years. Many pests are not only sporadic in time, but also occur sporadically throughout a field. The desirable approach would be to spot-treat those areas where the pest infestation exceeds the economic injury level. However, when spot treatment is not feasible, the pest manager must decide whether the potential losses from areas of the field economically infested would pay for the cost of treating the entire field. Generally in these situations, a farmer must learn to tolerate some damage in small areas in order to make money.

In the same way, the sporadic nature of certain pests over years must also be considered. It doesn't make economic sense to treat in "insurance fashion" a field that faces a low level loss in 1 year of 6 when the cost of the treatment over the years exceeds the value of the crop saved. An alternative would be to attempt to control the pest in those years when it appears. However, in some situations a small probability of high crop damage from pests exists. Such damage could seriously impair the financial stability of a farm operation. In this instance the application of insecticides, even though the added cost is higher than the potential gain, is the safe approach in terms of the total farm operation. It is important to consider the relevance of the nature of the pest population over years in making a pest control decision.

The economic injury level is the basis of most pest management decisions. Farmers are in business to make money and the use of unneeded pest control chemicals is a production cost that comes straight out of the profit column on the balance ledger.

Pest Management does indeed make "CENTS!"

Windshield May Be Clue

Bugs on the windshield! To most people that means poor driving vision and the need for strong windshield cleaner fluid. The pest management specialist may view the situation differently.

The presence of large numbers of insects on the windshield might signal an approaching pest problem. For instance, large numbers of moths on the windshield on a quiet night may mean that it's time to scout corn for corn borer eggs and larvae. The same might be true for moths of sod webworms or fall armyworms.

Entomologists also use light traps or pheromone traps to collect adult insects to determine the time for scouting operations. Philosophers still debate whether the chicken or the egg came first. Entomologists, on the other hand, know that after you see the moth the larva can't be far behind.

BUG SCOUT

"Thanks for the specimen, but your collection technique leaves something to be desired!"

Utilizing Heat Units

Insects are cold-blooded organisms! This means that the body temperature of an insect is close to the temperature in the environment. For this reason, biological functions of insects depend on temperature. Up to a point, insects move more quickly, eat more, and grow faster as temperature increases.

Pest management specialists use this basic fact to predict activity of insects. Heat units are calculated by summing temperatures above temperature thresholds for a specific time period. The threshold is the temperature where activity begins and is often referred to as the base. For example, it may take 670 heat units at base 50 for an insect to develop from egg to adult. Such information is helpful in timing scouting for insect damage.

Heat units are another tool utilized by pest management specialists to help farmers economically control insect pests.

BUG SCOUT

"That will teach you not to run around outdoors in the winter wearing those silly short pants!"

Pest Management

1. What is pest management?
2. Define economic injury level.
3. How would an occasional pest be different than a severe pest according to pest management concepts?
4. What are the two most important considerations that influence the economic injury level?
5. Why might it be said that pesticides, unlike fertilizers, don't increase crop yields?
6. How might heat units be used in insect management programs?
7. How might monitoring corn borer moth flights contribute to management of this insect?
8. What biological agent can be used for European corn borer control?
9. In addition to corn, what other crops might have European corn borer problems?

References Cited

Headley, J.C. 1972. Economics of pest control. Pages 180–187 in *Implementing practical pest management strategies.* Proceedings of a national extension workshop. Purdue University, Lafayette, IN.

Luckmann, W.H. and R.L. Metcalf. 1994. The pest management concept. Pages 1–34 in R.L. Metcalf and W.H. Luckmann, eds. *Introduction to Insect Pest Management.* Wiley, New York.

National Academy of Sciences. 1969. *Insect Pest Management and Control.* Washington, DC

Stern, V.M., R.F. Smith, R. Van den Bosch, and K.S. Hagen. 1959. The integrated control concept. *Hilgardia* 29(2):81.

Dealing with Insects That "Bug" Us

Household Pests

Ants (order: Hymenoptera)

APPEARANCE: Pest ants range in size from about 2 mm (pharaoh ant) to about 12 mm (black carpenter ant) in length. Most often we see wingless workers but reproductive forms have four wings. Unlike termites, ants have elbowed antennae, the second pair of wings is smaller than the first, and they have a slender waist.

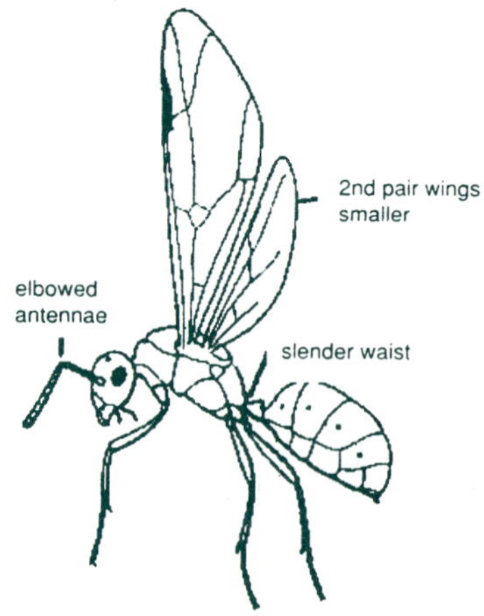

Ant

GENERAL INFORMATION: Ants cause problems by seeking food in our homes, constructing their nests in wood structures, producing mounds of soil in lawns and fields, and stinging humans and other animals.

WHAT TO DO: The best way to control problem ants is to locate and destroy the nest. This can be done with a direct application of an insecticide approved for that use or by placing insecticide baits so that food seeking ants carry the bait back to the nest.

Cockroaches (order: Dictyoptera)

APPEARANCE: Brownish black, brown or tan colored, flat bodied, ranging in size from about 12 mm to 38 mm in length.

GERMAN COCKROACH (12 mm) tan with two dark stripes on prothorax. Females carry egg case protruding from abdomen.

AMERICAN COCKROACH (38 mm) brown in color.

ORIENTAL COCKROACH (25 mm) black in color, females nearly wingless.

AUSTRALIAN COCKROACH (32 mm) brown in color, similar to the American cockroach except has a dark spot in the center of the pronotum.

BROWN-BANDED COCKROACH (10 mm) tan in color with light yellow crossband at base of wings.

WOODS COCKROACH (25 mm) brownish-black in color, male has wings and is similar in appearance to the American cockroach, female is wingless and similar in appearance to Oriental cockroaches.

ASIAN COCKROACH (12 mm) tan in color, similar to the German cockroach in size and color.

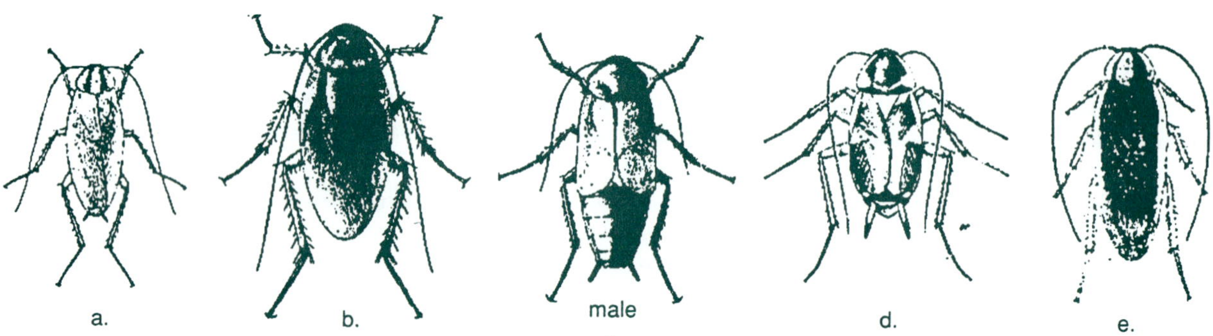

Cockroaches common to Indiana: a. German, b. American, c. Oriental, d. brown-banded, e. woods.

GENERAL INFORMATION: All cockroaches lay eggs in capsules (ootheca). Small roaches resemble adults but lack fully-developed wings. Development time from egg to adult ranges from as little as three months (German cockroach) to as much as 2.5 years (American cockroach). Most cockroaches are tropical insects and pest species survive in northern climates by living in heated structures. Most pest cockroaches in the United States are introduced species. There are many regional common names for cockroaches. For instance, in the South, the American cockroach is known as a palmetto bug, the Oriental cockroach is sometimes called a water bug, and on the East coast the German cockroach is sometimes called a croton bug.

WHAT TO DO: Use proper sanitation. Do not provide food such as crumbs behind the counter or in trash cans.
Minimize water sources from leaky faucets or condensation.
Reduce shelter by not allowing trash to build up and sealing cracks.
Guard against reinfestation by not bringing in infested material such as storage boxes from locations where cockroaches are present.
Use chemical treatment when necessary.

WHAT NOT TO DO: Ultrasonic devices and sticky traps. Sticky traps can be used to monitor populations but alone will not provide control.

Moths in Stored Food (order: Lepidoptera)

APPEARANCE: Dull colored moths with a wingspread of about 12 mm. There are three common species. The first is called the Indian meal moth. It has wings that are coppery colored near the ends and whitish at the base which makes the moth look banded when at rest. The Mediterranean flour moth has wings that are pale gray with black cross markings. A third species looks similar to clothes moths (buff colored, 10 mm in length) and is called the Angoumois grain moth.

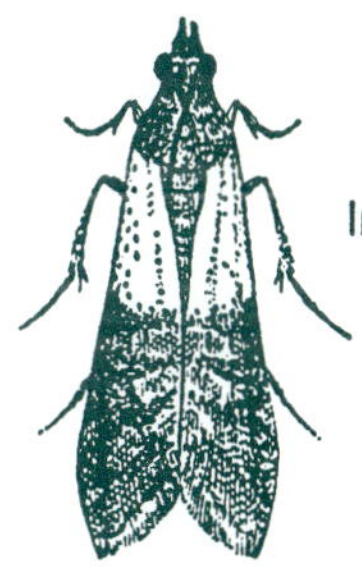

Indian Meal Moth

GENERAL INFORMATION: The immatures of both of these species are pinkish-white in color and web together the material where they are feeding. The immatures leave the food where they feed and crawl about cupboards and walls looking for places to pupate. The adults flutter about the house in the vicinity of the food source. The immatures of the Angoumois grain moth develop within kernels of grain such as popcorn or birdseed.

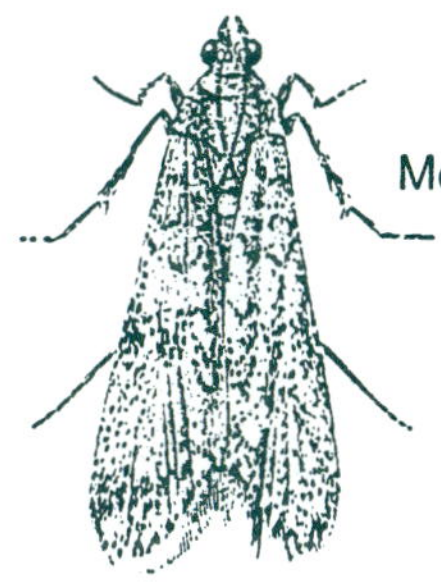

Mediterranean Flour Moth

WHAT TO DO: It is important to determine the source of the insects. Look for any plant-based material such as spices, flour, or pancake mixes. Destroy the infested material. To prevent infestations store susceptible material in insect-proof containers or the home freezer. When purchasing food that is susceptible make sure that the seals on the containers are not broken since many infestations begin in storage warehouses. Also keep food storage areas clean and free of crumbs or food particles. Ornaments and decorations that are infested can be placed in the freezer for 3 or 4 days to kill the insects.

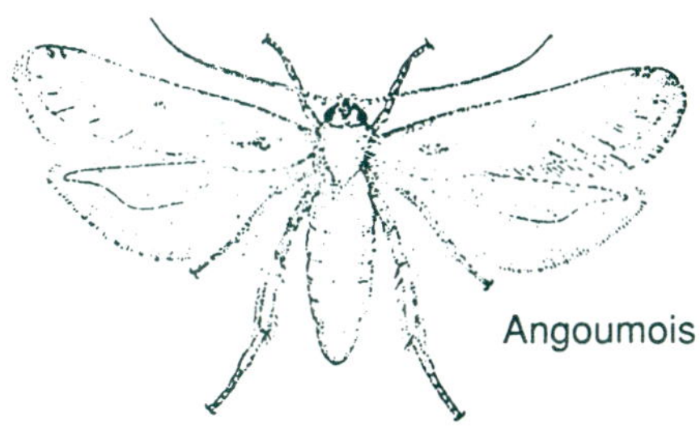

Angoumois Grain Moth

WHAT NOT TO DO: Ultrasonic devices will not help with insects in stored food.

Grain and Flour Beetles and Weevils (order: Coleoptera)

Granary Weevil

Rice Weevil

Appearance: There are several beetles that infest grain and flour and all are small (usually less than 3 mm long). The weevils are also small but have long snouts. The immatures are yellowish-brown and worm-like with brown heads.

General Information: The immatures are usually found in infested material. The beetles can be found in the infested food or crawling around the kitchen. Weevils primarily feed in whole grains. Their immatures are small, white legless grubs that develop inside individual kernels.

What to Do: As with grain-infesting moths, first identify the material that is the source of the insects. Remove any potential food sources and store material that is susceptible in insect-tight containers.

What Not to Do: Ultrasonic devices do not work for these insects.

Dermestid Beetles (order: Coleoptera)

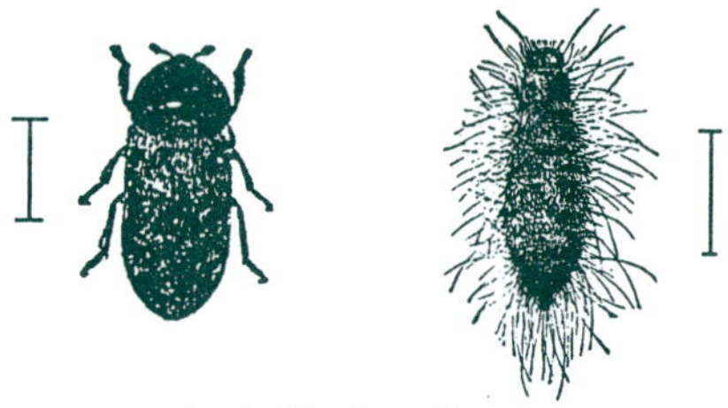

Larder Beetle and Larva

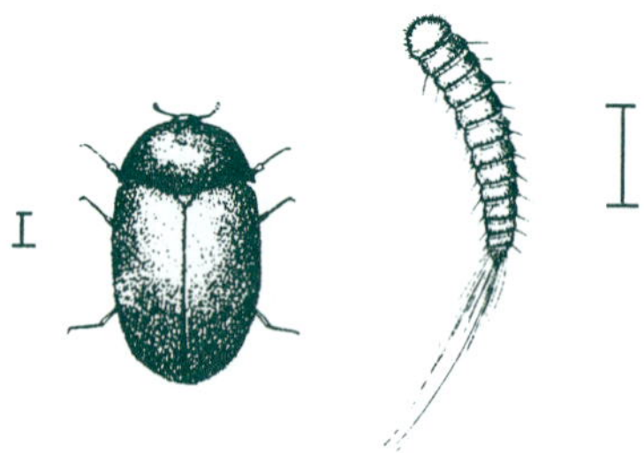

Black Carpet Beetle and Larva

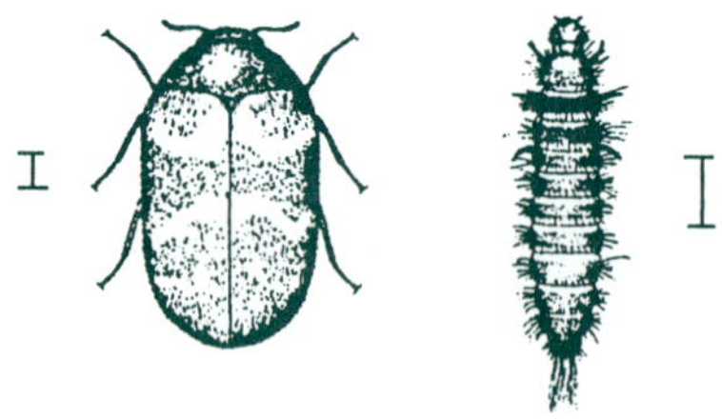

Cabinet Beetle and Larva

Appearance: These beetles are small (less than 12 mm in length), convex, dark-colored and hairy or scaly. The larvae are active and covered with long, dense dark hairs.

General Information: Dermestid beetles include insects called larder beetles, carpet beetles, and cabinet beetles. These insects are generally scavengers and feed on a great variety of plant and animal materials including leather, furs, skins, dried meat products, woolen and silk materials, cheese and cereal grain products. Larval stages of these insects do the most damage. Because the larvae are fuzzy they are sometimes called buffalo worms. Some of these beetles are referred to as museum beetles because they are used to consume the skin and hair from animals when the skeleton is destined for display.

What to Do: Destroy infested material and store items of clothing in insect-proof containers such as cedar chests or chests that contain moth-proofing materials such as moth balls.

What Not to Do: Ultrasonic devices do not work for these insects.

Clothes Moths (order: Lepidoptera)

Appearance: The moth is tiny (about 10 mm) and buff colored. The larvae is a small white caterpillar that lives in a silken case.

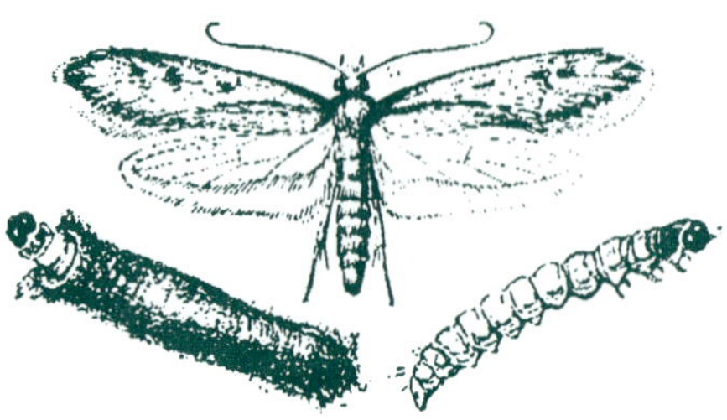

Casemaking Clothes Moth
(Top, adult; right, larva; left, larva encased)

General Information: Because the larva lives in a silken case it is called the casemaking clothes moth. It feeds on wool, hair, fur and feathers. The adult moth avoids light. You will notice these moths flying when the lights are turned on in a room. Otherwise they tend to rest on the walls in the vicinity of their caterpillar food sources.

What to Do: Remove and brush all susceptible garments each fall and spring. Hanging the garments in the sun (called sunning) for several hours will kill any larvae present. Dry cleaning or washing and pressing with a hot iron will also kill larvae. When the closet is empty vacuum and spray with an insecticide approved for this use. Clothing in storage closets can be protected with moth proofing chemicals such as moth balls or crystals. Infested rugs or carpets should be washed and sunned. Before replacing them, spray the floor and carpet with an insecticide approved for that use.

What Not to Do: Never store soiled garments because perspiration and food stains favor moth development. Ultrasonic devices also do not work for clothes moth control.

Ladybird Beetles (order: Coleoptera)

Appearance: One of the most recognizable beetles, ladybird beetles are small (1.5–6 mm) oval and convex beetles. They are often brightly colored and range from black with red spots to tan with no spots. Most are red or orange with black spots. Immatures are somewhat flattened and covered with spines. The immatures are also brightly colored and are usually banded or spotted.

General Information: Both adults and immatures are predators on soft-bodied insects, primarily aphids and scale insects. This makes the ladybird beetles valuable as biocontrol organisms. There are a couple of notable exceptions to this food habit, the Mexican bean beetle and the squash beetle which feed on plants. All ladybird beetles hibernate during the winter in groups sometimes in large numbers. This habit has made one species a nuisance pest. The multicolored Asian lady beetle was introduced into the US as a biocontrol agent for tree aphids. Now the insect invades homes in such high numbers seeking overwintering shelter that it has become a home pest as it finds its way from wall spaces and attics into living spaces during the winter months. Lady beetles do produce an odor, but they do not injure humans or pets or feed on anything in the house.

What to Do: It is impossible to keep all Asian lady beetles from finding a way into the house during the fall migration but caulking vertical and horizontal joints on sunny sides of the house will help. The best approach is to vacuum up those beetles that find their way into the living quarters of the house. Some very limited help might be gained by treating the sunny surfaces of the house with an insecticide when the beetles are seeking shelter in the fall.

What Not to Do: It generally does little good to spray the beetles in the house with an insecticide. This is because they will die quickly without food when they find their way into the house.

Termites (order: Isoptera)

Appearance: Most worker termites are small (about 12 mm), white colored, soft-bodied, wingless creatures. To distinguish from ants, termites have "fat" waists and straight antennae. Reproductive forms have four wings, similar in shape and size.

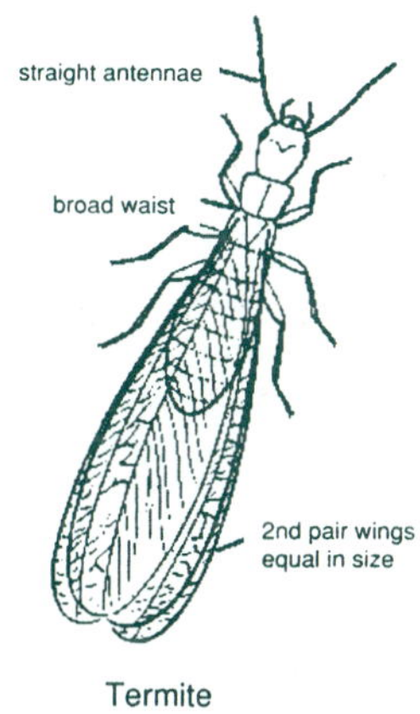

Termite

General Information: Termites are common in the warmer parts of the world and are most numerous in the tropics. In general there are two groups of termites. One type is where the nest is entirely or partly in the ground. In the other type the nest is entirely in wood. Most pest termites are of the ground nesting type. Termite food consists of wood, and paper or other wood products. Like other social insects termites have castes. Winged reproductive forms exist during the mating season. Symptoms of termite infestations include the following: pinpoint holes showing up in woodwork or wooden walls, bubbles or blisters in paint on wood, mud tubes connecting soil to the wooden part of a structure, or presence of winged reproductives in the structure.

What to Do: To prevent infestations do not have wood-to-soil contact associated with the structure (including decks). Eliminate excessive moisture anywhere in the structure including crawl spaces and in and around windows, doors and sinks. Do not pile wood against the structure. Once damage is present it is best to acquire the services of a pest control company for treatment. Termite bait stations that are placed in the lawn have proven effective for control.

What Not to Do: When termite damage is detected by you or reported by an inspecting pest control company never rush to contract for control services. Termites work slowly and it is always best to get a second opinion when considering a termite treatment.

Fleas (order: Siphonaptera)

Appearance: Adult fleas are hard skinned, spiney insects. They are wingless with back legs adapted for jumping. Fleas are thin from side to side.

General Information: There are many species of fleas and common names reflect a host animal such as the cat flea or dog flea. All fleas will feed on a wide range of fur-bearing animals and individual fleas will move from host to host to feed. Adult fleas feed on blood. Eggs are laid in the dirt or in the bed of the host. Flea larvae feed on dry organic matter including blood, feces or skin. Some fleas have a life cycle as long as 2 years but most complete development in 2–3 months. Some fleas remain in the pupal stage until an animal returns to the nest when the heat and vibrations stimulate the adult flea to emerge. Fleas are vectors of disease organisms including those that cause the plague and typhus.

What to Do: Flea control must include two aspects. First the fleas on the host animal must be destroyed. This can be done with a chemical treatment. The second aspect is to eliminate the larval population of the fleas. This means treating the pens or beds of the animal or the dry manure in barns. In homes the vacuum cleaner is a good tool to help combat fleas. Flea control is a continual process since it is easy for pets that have access to the outdoors to be reinfested with fleas from other animals.

What Not to Do: Do not use ultrasonic devices in the structure or on the pet. Do not have a pet treated without eliminating larval populations.

Lawn and Garden

Grubs (order: Coleoptera)

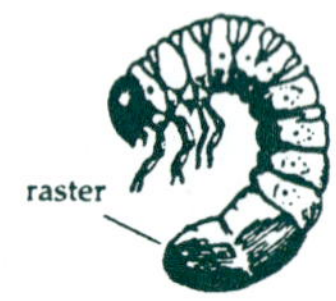

White grub

Appearance: White, C-shaped larvae with brown heads and 6 legs ranging from 5–40 mm in length depending on species and age. The rear end may appear dark colored.

General Information: White grubs are the immature stages of several species of beetles including the Japanese beetle, the masked chafer and the June beetles. These grubs feed on plant roots and begin feeding once they hatch. The grubs feed until the soil turns cold at which time they dig into the soil below the frost line to spend the winter. When the soil warms in the spring the grubs return to the root zone and, depending on species, either pupate or feed for another season. Feeding by white grubs kills the turf resulting in dead patches of grass in the lawn. Moles feed on white grubs and grub-infested lawns sometimes show heavy mole damage as well. Skunks, raccoons and black birds also feed on grubs and evidence of their feeding can sometimes show up on infested lawns.

What to Do: Chemicals can be applied to the lawn following egg laying by the beetles. This is generally around the first of August.

What Not to Do: Do not use light traps to reduce the number of beetles. The traps are not effective in reducing the number of grubs in the soil. The Japanese beetle traps also are not effective in reducing the number of grubs that end up in the soil.

Plum Curculio (order: Coleoptera)

Appearance: The plum curculio is a small (6 mm) dark-brown insect with whitish patches and four humps on the wing covers.

General Information: The plum curculio spends the winter as an adult and becomes active when apples and peaches bloom. The female cuts a small, crescent-shaped hole (a smiley face) in the small fruit and deposits an egg in the hole. The larvae eat their way through the fruit. Infested fruit are deformed because of oviposition and many fall from the tree.

What to Do: Removal and destruction of infested fruit will help by reducing the population of beetles. Jarring the tree with a stick during the time the beetles are active will cause the insect to drop into a cloth under the tree where they can be collected and eliminated. Insecticide sprays might be necessary during the maximum feeding and egg-laying period.

Japanese Beetles (order: Coleoptera)

Japanese beetle

Appearance: The Japanese beetle is metallic green and bronze, 8.5–15.9 mm long with reddish wing covers. It has several white spots near the tip of the abdomen.

General Information: The beetle feeds on the fruits and foliage of over 350 species of plants. The winter is spent as a grub (see section on white grubs) with most adults emerging in July. The insect was accidentally introduced into the US about 1916 in New Jersey on nursery rootstock imported from Japan. It is now present in most states east of the Mississippi. The insect is a major pest to the green industry and also causes damage to corn and soybeans.

What to Do: Plants, including fruits and vegetables, can be protected by spraying with an approved insecticide when the beetles appear. The treatment may have to be repeated as long as beetles continue to feed in the area.

What Not to Do: Japanese beetle traps using floral odors and pheromones are widely distributed at stores handling lawn and garden products. The traps collect beetles by the thousands but do not reduce damage to plants in the vicinity of the trap. To be effective such traps must be placed over 100 feet away from the area to be protected and must ring the area with placements every 50 feet. Light traps (bug zappers) also attract and kill many Japanese beetles but research has shown that plants in the vicinity of such traps sustain equivalent damage to similar plants not "protected" by light traps.

Mosquitoes (order: Diptera)

Appearance: Mosquitoes are small (less than 12 mm in length), rather fragile looking insects with two membranous wings fringed with scales. They have slender mouthparts and the males have bushy antennae.

Important Information: Adult female mosquitoes require a blood meal to allow them to lay eggs. Because of this blood-feeding habit, mosquitoes are problem insects for several reasons. Worldwide there are over 2,500 species of mosquitoes with about 140 species in North America. Some transmit a number of diseases to humans and animals. Among the human diseases are malaria, yellow fever, and viruses that cause encephalitis. A new mosquito-transmitted disease in the US is West Nile Virus. The major reason that most people are concerned with mosquitoes is because of the annoyance and irritation of the bites. Immature mosquitoes develop in water. Depending on the species that water could be located in floodwater pools, ditches, marshes, tree cavities, rain barrels, birdbaths, old tires, or gutters. The larvae are called wigglers. Mosquitoes can go from egg to adult in as little as 10 days so populations can build up rapidly during periods of wet weather.

What to Do: To prevent mosquito bites the best way is to use repellents. Repellents are available under a variety of trade names. Such repellents will provide protection for up to 5 hours depending on perspiration, temperature and abundance of mosquitoes. Follow directions on the label carefully when using repellents. Mosquito populations around the home can be reduced by eliminating breeding sites. In other words getting rid of free standing water. In ponds top-feeding fish can be introduced to help reduce immature populations. Also change the water in birdbaths or wading pools every week. Not allowing weeds to grow uncontrolled around the home or adjacent lots will reduce daytime resting sites for adult mosquitoes.

WHAT NOT TO DO: Light traps do not work to reduce biting mosquito populations. Some male mosquitoes are attracted to light and are killed in light traps but they do not bite. Also, ultrasonic devices do not provide protection against biting mosquitoes. Citronella candles are of limited value in reducing mosquito bites. The same is true for plants called "mosquito plants."

Bees and Wasps (order: Hymenoptera)

APPEARANCE: Bees and wasps that sting are fair-sized (12–25 mm) insects. They are generally yellow and black or brownish in color, frequently with stripes. Bees generally are robust in build with hairy bodies, while wasps are more slender and lack fuzz.

GENERAL INFORMATION: Some bees and wasps sting in defense of themselves or their nests. Honey bees can sting only once because their stinger stays in the skin of the victim. Other bees and wasps can sting multiple times. The toxin injected in a bee sting is a complex chemical mixture that disrupts nerve transmission and breaks down cells. The response of most people to the toxin is a localized reaction including swelling, pain and redness. A few people, however, react to a bee sting with an allergic reaction that can be life threatening.

WHAT TO DO: If stung by a honey bee remove the stinger as quickly as possible. Wash the area carefully with soap and water. Apply a cold pack to the affected area. Apply a paste of baking soda and water and leave it on for 15 minutes. Take pain killers as needed. For itching dab on a tiny amount of household ammonia. Take an antihistamine if approved by your physician. You should seek medical attention under these conditions: large areas of swelling, abnormal breathing, tightness in throat or chest, dizziness, hives, fainting, nausea or vomiting, or persistent pain or swelling.

WHAT NOT TO DO: In general leave bees and their nests alone. Do not swat at bees or wasps that fly around you.

Aphids (order: Homoptera)

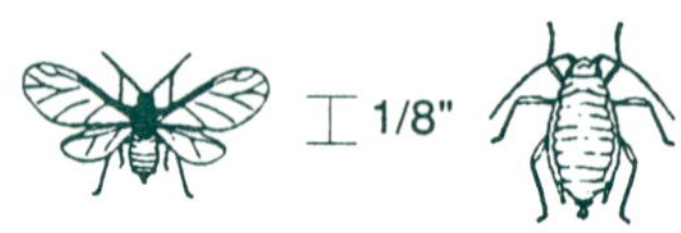

Aphids

APPEARANCE: Aphids are small (2 mm) soft-bodied insects. All have a pair of protrusions on the abdomen called cornicles.

GENERAL INFORMATION: These insects are sometimes called plant lice. Aphids are pests because they suck the sap as food and stress the plants. In addition the saliva of some aphids is toxic to the plant and a few aphids transmit viral diseases. Aphids can have several generations per year. There are hundreds of species of aphids and all have a complex life cycle many times moving between types of plants during the year. There are winged and wingless forms and many females give birth to live young and reproduce without mating. Aphids are food for several predators including ladybugs and lacewings. There are also a number of insects that are parasitic on aphids.

WHAT TO DO: Make certain the aphid population is damaging and evaluate the presence of biocontrol insects such as ladybugs. Choose chemicals that have the least hazard to predators and parasites. When chemicals are needed several are available including dormant oils where the target aphid overwinters in the egg stage.

WHAT NOT TO DO: In general predators will control many aphid populations and parasites, so it is important not to disturb the natural system with chemicals unless necessary.

Field Crops

Boll Weevil (order: Coleoptera)

APPEARANCE: The adult is small (6.4 mm) and hard shelled. The immature is a curved white grub 12.7 mm in length.

General Information: The cotton boll weevil is one of the most recognizable insect pests in the world. It is a native of Mexico and first appeared in the US in 1892. Its presence has caused economic impact beyond damage to the cotton crop including land values and the closing of cotton gins and oil mills. Damage is caused by the adult weevils and their grubs. The adults chew holes in the squares and bolls of the plant. Eggs are placed in the holes and the grubs feed in the cotton boll. This causes the boll to drop off or produce less fiber. The boll weevil spends the winter in the adult stage hiding in protected places such as plant debris. The male sex pheromone called "grandlure" is used to monitor and trap boll weevil populations.

What to Do: In general control of boll weevil falls into three categories. The first is destruction of plants in the fall following harvest. The plants harbor boll weevils that will overwinter. It is important to plant the crop early and get the plants beyond the most susceptible stage before the weevils emerge to lay eggs. Finally, it may be necessary to use chemicals when the weevil threshold is exceeded.

Corn Rootworms (order: Coleoptera)

Appearance: There are three common species of corn rootworm beetles and all are 5–6 mm in length. The northern is uniformly green to yellowish in color. The western is yellow-brown with three prominent black stripes. The southern is green with 12 black spots. Immatures are white in color with a brown head capsule and a brown spot on the rear.

General Information: Corn rootworms are beetles native to North America that have larvae that feed on corn roots which is the basis for their name. There are three common species that damage corn in the US. The western and the northern overwinter in the soil as eggs. The southern adults fly into the Midwest each spring. The feeding of the immatures destroys corn root tissue which interferes with the ability of the plant to procure water and nutrients. In addition the root damaged plant tends to fall down which further reduces the ability of the plant to produce an ear. Crop rotation (where corn is planted following other crops) has traditionally been used to eliminate damage by the northern and western. A change in the egg-laying habits of the western adults coupled with cropping technology has reduced the effectiveness of this approach in some areas in recent years. The western corn rootworm has been accidentally introduced into Europe.

What to Do: Rotate crops so that corn is not grown on the same land for two consecutive years. Monitor fields for corn rootworm beetles to assess if the immature populations will be damaging the next year. Apply insecticides to the soil during planting to kill larvae before they damage the crop.

European Corn Borer (order: Lepidoptera)

Appearance: Adult borers are moths that are pale yellowish brown with dark wavy lines. The wings when extended are about 25 mm across. When at rest, such as on lighted window panes the shape of the moth is a triangle. The larvae are a flesh color with inconspicuous brown spots.

General Information: The European corn borer is a moth that is native to Europe and Asia and has been in North America since at least 1917. The insect arrived in shipments of broomcorn from Italy or Hungary. Immatures bore into plants, hence its name. It feeds on many species of plants but where corn is grown it feeds almost exclusively on that crop. The insect passes the winter as a mature larvae in the stalk of the plant where it fed. The adults emerge in early summer and fly to attractive plants for egg laying. The larvae first feed on the surface of the leaves and then bore into the plant. The feeding of the larvae reduces the ability of the plant to translocate water, minerals and photosynthetic products within the plant and ultimately reduces yield. In addition the feeding of the larvae weakens the stalk making it prone to breakage and susceptible to disease organisms. The number of generations of borers depends on the area and may be as high as three in southern regions of the US. Second generation borers attack the ears of corn and are sometimes found in the ears of sweet corn.

What to Do: Corn borer larvae can be controlled by applications of insecticide to the corn before the larvae begin to bore into the plant. Lines of corn have been developed that resist the feeding of newly hatched larvae, so when these lines are planted the damage by the borers is avoided or reduced. In addition, through genetic modification, the genes that produce the protein in the bacteria BT that is toxic to Lepidoptera larvae has now been incorporated into the corn plant. Called BT corn, larvae that try to feed on the plant are killed by the toxin. Farmers who are faced with a high infestation of corn borers will try to harvest the crop as soon as possible to reduce the losses due to ears dropping to the ground.

WHAT NOT TO DO: Programs in the 1920's and 1930's where stalks harboring the borers in the overwintering stage were destroyed by plowing under the soil or burning proved ineffective. This was primarily because enough borers survived in alternate hosts to create problems in the following year.

Hessian Fly (order: Diptera)

APPEARANCE: The fly is small (3.2 mm) sooty-black color. It is smaller than common mosquitoes. The immatures are small, white maggots.

GENERAL INFORMATION: The Hessian fly is one of the old pests in the US. It is a pest of cereal grains where the maggot feeds in the stem of the plant. The original home of the Hessian fly was Russia. It arrived in the US in straw bedding used by the Hessian troops during the Revolutionary War and was first reported in 1779. The Hessian fly passes the winter in the full-grown maggot stage in the wheat stem. The flies lay eggs on the plant in the fall.

WHAT TO DO: Two approaches to controlling the Hessian fly have been practiced. First the use of a planting date to avoid egg laying by the adult fly. The crop is planted after the date when the last flies have died. This date is known as the fly-free date. The date varies from north to south in the country. For example the fly-free date in southern Michigan is September 20 while the date in southern Illinois is October 12. The second method is to plant lines of wheat that do not allow the larvae to survive, termed resistant lines. Such lines have been developed by scientists of the US Department of Agriculture and have been widely used to control this insect.

WHAT NOT TO DO: With the availability of resistant lines of wheat farmers tended to not use the fly-free date for planting. Such widespread exposure of resistant lines of wheat to large numbers of flies increased the probability for development of flies that survived on the resistant lines. The result was that the resistant lines more quickly became ineffective for fly control.

Control in Home and Garden

1. Why will insect control gadgets such as light traps not provide satisfactory control?
2. Why is sunning considered a method of control for clothes moths or carpet beetles?
3. What are three physical characteristics that will distinguish ants from termites?
4. What is the only ant that will cause structural damage to houses?
5. What is the absolute best (the big number one) way to prevent ants from invading a house?
6. Name six species of cockroaches likely to be found in homes.
7. The chances of successful roach control are greatly increased if what precedes the application?
8. Wiggle-tails are the larvae of what insects?
9. What is the most effective control procedure for mosquitoes around the home?
10. Name five groups of insects that are considered pests of home-stored foods.
11. List five procedures useful in preventing infestations of stored foods.
12. What are symptoms of grub infestations in lawns?
13. What are the common names of three insects with larvae known as white grubs.

CHAPTER 21

Entomophobia

Who's Afraid of the Big, Bad Bugs?

The ancient Greeks used the word "entomon" to mean insects and almost anything creepy or crawly, which likely included snakes, lizards, frogs, toads and worms of all sorts. Today, we use entomo- as a combining form to mean anything dealing with insects. For example, entomology is the study of insects. And a person who studies insects is known as an entomologist.

Not everyone finds insects as interesting as entomologists do. In fact, some people react quite differently—that is, with an extreme fear of insects. These folks have a disease known as entomophobia.

A phobia is an unfounded fear of something, and entomophobia is an unfounded fear of insects. Estimates indicate that nearly 10 percent of the U.S. population suffers from this disease.

Phobias are interesting diseases in that medical science cannot explain what causes them. Certainly, entomophobia is not a contagious disease—it is not caused by some disease organism that can be transferred from person to person. Nor is it likely that the disease is hereditary, passed from parents to offspring through their genes.

Most entomologists agree that entomophobia is learned. We learn to fear the little six-legged creatures because of the behavior and advice of those around us.

Most insects pose little danger to humans. Some, however, can be hazardous to human health. It makes sense to be somewhat wary of bees and wasps. Avoiding biting flies and mosquitoes might prevent the discomfort these insects can inflict.

If there is a real basis for entomophobia, it might be related to the role that insects play in disease transmission. Insect-borne diseases, including malaria, sleeping sickness and typhus, have caused untold human suffering and death. Consequently, in ancient times it might have been logical to fear the insects that were purveyors of death. Since many insects are difficult to tell apart, except maybe to entomologists, it could have been a matter of survival to fear them all.

This attitude is, no doubt, supported today by the reality that some insects, such as some flies and cockroaches, prosper under less-than-sanitary conditions. No one wants to be associated with such filth-loving organisms.

Never mind that the vast majority of insects are not harmful to humans and don't even live in filth. Most insects are pronounced guilty because of their "cousins."

Entomophobia may prevent some people from making friends with insects. But insects probably won't shy away from people. After all, many insects don't suffer from anthrophobia—the fear of humans.

Scratch When It Itches

The thought of insects crawling on humans make some people itch. Thanks to modern technology, having insects on our persons is not as common as decades ago. But we still worry about the prospects of such intimate intruders. Thoughts of lice, fleas and mites still crawl through our heads as their ancestors did on our ancestors in years gone by.

So creative is the human mind, and so real is the sensation of sharing our bodies with guests of the insect kind that some people actually conjure up a personal infestation of little "beasties." Imagining that insects are crawling on our bodies is a disease known as delusory parasitosis.

"Delusory" means something really doesn't exist. "Parasitosis" is the medical term for having parasites on the skin. Delusory parasitosis differs from other hallucinations in that it is a tactile sensation. Other hallucinations cannot be felt.

People suffering from delusory parasitosis frequently suffer in silence. Society today, unlike in

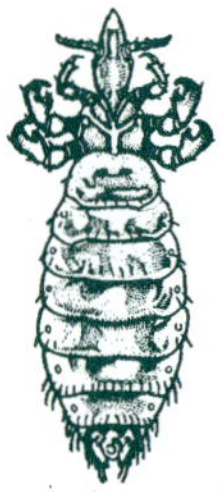

ancient times when it was considered fashionable to harbor a personal crop of insects, frowns on such infestations. To admit to being infested is certain to elicit raised eyebrows if not downright scorn.

Such folks frequently take drastic measures to deal with the problem. They have been known to spray their homes daily with insecticides. This normally does not work since insecticides cannot kill imaginary insects.

The solution to overcoming this phobia involves working with a trained therapist who is aware of this disease. In this case, an entomologist isn't prepared to deal with what "bugs" people suffering from delusory parasitosis.

A "Bug" Phobia

Panic sets in as a large dot, which you thought was part of the wallpaper design, begins to noticeably move toward you. Your heart pounds loudly in your chest as you realize that the dot is at least an inch long, and has six, ugly legs. How dare this nasty insect invade your home. You then swat it with a newspaper and declare victory over the invader.

Why is it that we humans fear these little creatures? The vast majority of insects are harmless to people although some can sting and bite. Therefore, most insects need not be feared. However, a high percentage of the human population approaches interactions with the insect world on the premise that they never met an insect they didn't hate. It has been estimated that nearly 10 percent of us big, brave humans show an irrational fear of insects, a condition known as "entomophobia."

Scientists have long debated the basis for entomophobia in the human population. It has been suggested that there is a biological tendency for this phobia based on the potential harm some insects can cause such as stinging or even transmission of diseases. Entomophobia may be an extension of generalized animal phobias, where children are said to first enjoy and then develop a phobia for animals in their environments.

In the book *Dragons of Eden*, Carl Sagan has suggested that a fear of insects might be a modern remnant of early human conflict with large reptiles. Regardless of the basis for this irrational fear of insects, it is clear that this is a learned behavior and our experiences modify our attitudes toward insects. There is no question that the onset of entomophobia occurs in childhood. In general, fear of insects is more common among children than adults and greater among women than men, although it may be that men are more unwilling than women to admit that they are afraid of a little "bug."

Even the syndicated columnist Ann Landers responded to letters from entomophobics. Ann's advice in a September 3, 1980 column was "to read up on bugs and become informed. They are fascinating creatures." Indeed education is the key to overcoming most fears, including entomophobia.

Whether or not education would have worked for an entomophobic young lady by the name of Muffet is unclear.

Little Miss Muffet
Sat on a Tuffet,
Eating of curds and whey;
There came a big spider,
And sat down beside her,
And frightened Miss Muffet away.

—Mother Goose, 1916

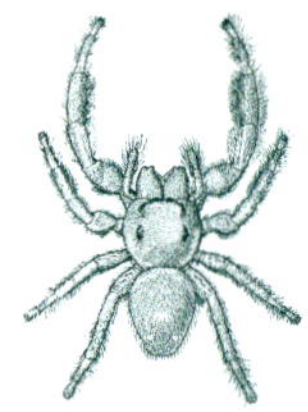

Even though a spider is not technically an insect (it has too many legs) Miss Muffet's response is a common one when coming face to face with an insect.

Entomophobia

1. What is a phobia?
2. What percentage of the U.S. population suffers from entomophobia?
3. What is delusory parasitosis?
4. How does delusory parasitosis differ from other hallucinations?
5. What is the key to overcoming most phobias including entomophobia?

CHAPTER 22

Insects and Human Culture

Insects in Music

Jim Crack Corn (The Blue Tail Fly)

When I was young I used to wait
On master and hand him the plate:
Pass down the bottle when he got dry,
And brush away the blue-tail fly.

Chorus:
Jim crack corn, I don't care,
Jim crack corn, I don't care,
Jim crack corn, I don't care,
Ol' Master's gone away.

Then after dinner master sleep,
He bid this fellow vigil keep;
And as he's 'bout to shut his eye,
He tells me watch the blue-tail fly.

An when he rides in the afternoon,
I follow with a hickory broom;
The pony being very shy
When bitten by the blue-tail fly.

One day he rode around the farm.
The flies so numerous they did swarm;
One chanced to bite him on the thigh,
The devil take the blue-tail fly!

The pony ran, he jump and pitch,
And tumbled master in the ditch
He died and the jury wondered why,
The verdict was: the blue-tailed fly.

They laid him under a 'simmon tree,
His epitaph is there to see;
"Beneath this stone I'm forced to lie,
All because of the blue-tail fly."

Old master's gone, now let him rest,
They say all things are for the best;
I'll never forget, till the day I die,
Old master and the blue-tail fly.

Shoo, Fly, Don't Bother Me

Chorus:
Shoo, fly, don't bother me!
Shoo, fly, don't bother me!
Shoo, fly, don't bother me!
I belong to Company G.

Verse:
I feel, I feel, I feel,
I feel like a morning star.
I feel, I feel, I feel,
I feel like a morning star.

Charlie Caterpillar

Look at all the bees and ants standing all around,
Making fun of Charlie crawling on the ground.
"But wait!" cried Charlie.
"Wait and see,
For next week a butterfly I'll be!"

(spoken) Uh-huh!

La Cucaracha

La cucaracha, la cucaracha, ya no quiere caminar.
Porque no tiene, porque le falta marijuana que fumar.

Ya se van los Carranzistas,
Ya se van para Laredo.
Ya no son Convencionistas,
Porque tienen mucho miedo.

The Black Fly Song

'Twas early in the spring when I decide to go,
For to work up in the woods in North Ontario;
And the unemployment office said they'd send me through
To the little Abitibi with the survey crew.

Chorus:
And the black flies, the little black flies,
Always the black fly no matter where you go;
I'll die with the black fly a pickin' my bones
In North Ontari-o-i-o, in North Ontario.

Now the man Black Toby was the captain of the crew.
And he said: "I'm gonna tell you boys what we're gonna do.
They want to build a power dam and we must find a way
for to make the Little Ab flow around the other way."

Chorus: With the black flies . . .

So we survey to the east and we survey to the west,
And we couldn't make our minds up how to do it best.
Little Ab, Little Ab, what shall I do?
For I'm all but goin' crazy on the survey crew.

Chorus: Oh, the black flies . . .

It was black fly, black fly, black fly everywhere,
A-crawlin' in your whiskers, a-crawlin in your hair;
A-swimmin' in the soup and a-swimmin' in the tea;
Oh the devil take the black fly and let me be.

Chorus: And the black flies . . .

Black Toby fell to swearin' cuz the work went slow,
And the state of our morale was a-getting pretty low,
And the flies swarmed heavy; it was hard to catch a breath
As you staggered up and down the trail talkin' to yourself.

Chorus: And the black flies . . .

Now the bull cook's name was Blind River Joe;
If it hadn't been for him we'd've never pulled through.
For he bound up our bruises and he kidded us for fun,
And he lathered us with bacon grease and balsam gum.

Chorus: For the black flies . . .

At last the job was over; Black Toby said, "We're through
With the Little Abitibi and the survey crew."
'Twas a wonderful experience and this I know—
I'll never go again to North Ontario.

Chorus: And the black flies . . .

The Boll Weevil

The first time I seen the Boll Weevil
He was a sitting on the square.
The next time I seen the Boll Weevil
He had this whole darn family there,
Just looking for a home; just looking for a home.

The Boll Weevil said to his loving wife,
"Honey, stand up on your feet;
Look over yonder in Arkansaw
At all the cotton we've got to eat.
We'll have a home; we'll have a home."

The Boll Weevil said to the lightning bug,
"Can I get up a deal with you?
If I was a lightning bug,
I'd be working the whole night through,
I'd have a home; I'd have a home."

The Boll Weevil ate half the cotton,
And the banker got the rest;
Didn't leave that farmer's wife
But one old cotton dress,
And it's full of holes, and it's full of holes.

Excerpts from Praying Mantis

(Don Dixon)

Chorus:
"I feel like a praying mantis, her scent's around me
I feel like a praying mantis, I sense your antics
But I can't help it, I've been too frantic too long.

ROCK 'N' ROLL ARTIST NAMES INFESTED WITH INSECTS

In 1979, historian Carson Ritchie wrote a book about insects and human history entitled "Insects, The Creeping Conquerors." The book chronicles the effect that insects have had on people. It is considerable.

But Ritchie concedes he omitted, "in order to keep the book within readable bounds," a discussion of how insects have influenced human music. Indeed, in human music as well as in the rest of human history, insects are there!

Take rock 'n' roll for instance. Beginning with the Beatles, insect names for rock 'n' roll artists are as common as, well, fleas on dogs! The Beatles based their name on the "beat" music they were playing, and the most common type of animal on Earth—beetles. They wanted an insect name because they admired the U.S. group known as Buddy Holly and the Crickets. Imitation is the sincerest form of flattery!

The history of rock 'n' roll is infested with artists who have chosen insect names. Even the word insect shows up—as in the group called Insects. Sometimes the word insect is combined with other words to form a name for a rock 'n' roll group. For instance, Insect Funeral. We also find groups called Insect Jazz and Insect Opera, apparently in reference to the songs that real insects produce. There are also the Insect Surfers and Startled Insects.

Not surprisingly, insects that sing are popular in rock 'n' roll artist names. Such a singing insect and a group is Katydids. Crickets have been referenced as merry in poetry, and we find a rock group named Happycrickets. If one singing insect is good, two must be better: thus the name Grasshopper and Golden Crickets. In addition to their singing prowess, grasshoppers are also known for their numbers, as is referenced in the groups Grasshopper Highway and Grasshopper Takeover.

The locust, which is an incorrect common name used for cicadas in North America and correct for grasshoppers in the rest of the world, also shows up in rock 'n' roll names. We find groups called Chrome Locust, Hungry Locust, Locust Fudge and Distant Locust.

Flies are naturals to be included in names for rock 'n' roll artists. There are real types of flies, including Horse Flies, Domestic Flies and Tse Tse Fly. Other flies include Twenty Ton Fly, Lounge Fly, Madfly and Milky Fly. There is also Flyscreen, Flyspeck and Fly Swatter. Even immature flies get in on the act in the names Maggots from Mars and Baby Flies. There are rock groups called Four Flies on Grey Velvet and Fly Ashtray.

Bees also abound in rock 'n' roll group names. Bee names include Busy Bee, Killer Bee, Dance Bee, Sick Bees and Sugar Bees. The success of Buddy Holly and the Crickets was probably the inspiration for the name Honey and the Bees. An insect cousin to the bees shows up in another Buddy Holly name mimic, Chico and the Hornets. Bees and hornets are both stinging insects. The result of such a sting is included in the name of Bee Stung Lips, a reference to the full, red pouty lips that were a fashion statement back in the flapper era.

Buddy Holly would be proud of other insect names as well—Freddie and the Fleas, for instance. But that is not the only rock group with flea in its name. Consider Atomic Flea, Beach Flea and Fleaboy as other examples. There is also a group that uses an ancient flea control technique in its name, Saturn's Flea Collar. Speaking of insect control names, there are also groups called Roach Motel Style and Roachpowder. The latter is an obvious reference to an insecticide for these pest insects.

One of the first popular hard rock groups had the name Iron Butterfly, an obvious play on the delicate nature of this group of insects. Butterflies show up in many rock group names, including Butterfly Temple, Butterfly Train and Butterfly Messiah. The group Butterfly Tree took its name from the roosting trees where migrating monarch butterflies land to spend the night. Butterfly Child is the name of a rock group. However, the term is also used to describe a genetic human disease where the skin lacks anchoring fibrils, creating a condition where the child is "as fragile as a butterfly."

All sorts of insects are included in rock artists names. Adam Ant or Adam and the Ants and Alien Ant Farm represent these sometimes pesty creatures. There is also a group called Termites.

The light-producing beetles known as fireflies are the basis for rock group names as well. There is a group called Firefly and another called Fire Fly. Since one word is the correct spelling for the light-producing beetle, the use of two words suggests that the group has the name of a real fly associated with fire. The immature firefly is included in the name Glo-worm. And the Buddy Holly model shows up again in Kory and the Fireflies.

In the history of rock 'n' roll, insects are there, and it all started with a couple of crickets chirping in Buddy Holly's garage!

"She had insect eyes, but I could still see
That the look she gave him, you gave to me"

She bit off his head so he would not feel the pain
She wanted his body so much she ate his brain
Yeah, but I'll survive. No you won't catch me.
I'll resist the urge that is tempting me
I'll avert my eyes, keep you off my knee
But it feel so good when you talk to me."

***Excerpts from* Glowworm**
(The Mills Brothers)

Glow little glowworm, fly of fire,
Glow like an incandescent wire.
Glow for the female of the species,
Turn on the AC and the DC.

This life could use a little brightin'
Light up you little ol' bug of lightnin'.
When you gotta glow, you gotta glow!

Not every nautical boll weevil,
Illuminates yon woods primeval.

Glow little glowworm, turn the key on,
You are equipped with a taillight neon.
You've got a cute vest, sparkin' master,
Which you can make both slow and faster.

Some Well-Turned Phrases Insect Infested

A concise or aptly expressed thought is sometimes captured in what can be called a well-turned phrase. Such word combinations are memorable. They stand out from everyday language.

According to Purdue linguistics professor Victor Raskin, some well-turned phrases find common usage in language. When that happens, the phrases sometimes become trite and are classified as clichés.

Throughout history, some of the finest wordsmiths have turned to the insect world to produce well-turned phrases. Homer the Greek poet did in his epics. To describe the Greek host at Troy, Homer wrote "as numerous as flies near the farmer's milk in spring time."

Homer described troops pouring out to battle in the Trojan War as "wrathful wasps teased and roused to action by irresponsible young boys." This word picture in living color is as descriptive today as it was in Homer's time.

The poetic language of the Bible provides a well-turned phrase or two with insects. From King Solomon we learn: "Go to the ant, O sluggard; consider her ways, and be wise." And in an effort to express a great number, this phrase comes from Judges 6-5: ". . . and they came as grasshoppers for multitude for both they and their camels were without numbers."

When it came to crafting a well-turned phrase, no one has done it better than Shakespeare. In "Henry VI" we learn, "The smallest worm will turn, being trodden on." In "Love's Labor," Shakespeare describes a minimal amount as ". . . have no more man's blood in's belly than will sup a flea."

And speaking of fleas, in "Henry VII" we read: "That's a valiant flea that dare eat his breakfast on the lip of a lion." From fleas to flies, the Bard describes the unromantic Romeo in this way: ". . . more courtship lives in carrion flies than Romeo, . . ." In "King Lear" we read: "As flies to wanton boys are we to the gods. They kill us for their sport."

Modern writers also turn a phrase or two with an insect theme. To describe her disdain for insects, Pulitzer Prize winner Annie Dillard writes the following in "Pilgrim at Tinker Creek": "More than one insect—the possibility of fertile reproduction—is an assault on all human value, all hope of a reasonable god." Dillard also describes termite workers carrying eggs away from their queen as "longshoremen unloading the Queen Mary."

Dayton Hyde wrote a nature story about a farmer raising a litter of coyote pups entitled "Don Coyote." In it, he describes a coyote pup carrying a petunia blossom ". . . as though it were a brilliant butterfly . . ." It's easy to visualize how the pup was carrying the flower!

Kobo Abe in "The Woman in the Dunes" describes a passionate woman as "glowing from within now, as if she were being washed by a wave of fireflies." And speaking of fireflies, it was astronaut John Glenn who described the sight from the space capsule in his first orbital flight as "a field of fireflies."

And on a less-flattering level, a few years back in the wake of the scandal surrounding the International Olympic Committee, some of the principles were described as "running like cockroaches from the light." It may not be a well-turned phrase, but hardly a person is alive who doesn't understand what is being said.

So it is with other sayings. A fly in the soup or the ointment is generally understood. And everyone knows what is happening when a bug is put in someone's ear. While these insect sayings might at one time have been well-turned phrases, time and usage have turned them into clichés.

Insects Creep into Everyday Language

Insects, it seems, are everywhere. Even our language is crawling with insect references. We have been known to describe each other in insect terms.

Almost everyone has heard a child depicted as "cute as a bug's ear!" I've always assumed that such a comparison to a bug's ear was a positive reflection. But how cute is a bug's ear anyway? It's hard to say. You see, insects don't have ears!

We assume that being as cute as something that doesn't exist is really cute. So cute, that it is impossible to describe—as indescribable as an old marketing slogan once proclaimed a delicious taste was.

A person with bushy, protruding eyebrows could be described as beetle-browed. Such a description is a double-whammy. The origin of the term beetle literally means to project, which is exactly what a brow is designed to do.

Eyes can also be described in insect terms, as in bug-eyed; of course that means that the eyes protrude more than normal. This is frequently the case in insects, where the eyes are the prominent feature of the head. Protruding eyes allow an insect to see in all directions without turning its head. Humans have the capacity to "bug" out their eyes in a show of emotion.

A person who is slight of build is sometimes referred to as a flyweight. The term is even used for a weight class in boxing. Flyweight is also used as a demeaning reference to someone who does not come up to some standard.

Beetlehead is a term used for "stupid fellow" or someone who does something really dumb. The word beetle historically has been used for a wooden hammer. Thus, beetlehead is used interchangeably with hammerhead. Neither is a term of endearment. The comic strip character Beetle Bailey no doubt is named because he does some really dumb things!

We also describe our behavior in insect-related terms. Someone with "ants in their pants" is fidgety and can't sit still. Of course, anyone who knows anything about ants is aware that these insects bite. Some, such as fire ants, sting. Either way, if ants are crawling on you, there is a chance of being bitten or stung. That is enough to make anyone wiggle around a little.

We even use the word "antsy" to describe a person who is anxious about something. While we don't go so far as to say that the person has ants on his or her person, the term suggests that might be so.

People who might be considered overzealous about something could be described as having a bee in their bonnet. If you've ever had a bee in your chapeau or shirt for that matter—you know that the insect becomes an issue of utmost importance. It is so important that the afflicted person does not rest until the insect has been dispatched—thus, the accuracy of the description.

Sometimes a person is said to be busy as a bee. We all know that bees are hardworking. They have to be industrious to gather all of that nectar to make the winter's supply of honey. So a hard-working person is compared to the bee. But how much does a bee work? Not as much as most of us assume. Research has shown that most honey bees spend as much as 70 percent of their time loafing! But poet Ogden Nash says that we believe bees are hardworking because they are smart enough to buzz!

We also refer to making a beeline for something. The assumption is that we follow a straight line, as does the bee. Well, bees do take the most direct route to flowers when they follow directions transferred from another bee during the famous bee dance. However, after they have traveled the route, they navigate by sight and may detour around trees and structures—so much for making a beeline to something. That beeline often is not very straight!

Having insect ectoparasites, such as fleas and lice, has been a way of life for humans throughout history. Only recently have we been able to live without having to feed these insect moochers. But we still describe ourselves in terms related to these insects. If you aren't feeling very well, you might describe yourself as feeling lousy. In other words, you have a good infestation of lice.

Sometimes an unhealthy-looking person is described as flea bitten. This is an indication that he or she is being drawn down by an excess population of these blood-sucking insects.

In a less serious vein, getting nervous before some event is sometimes put in terms of "having butterflies in your stomach." That, we must assume, is a step up from feeling lousy!

Excerpts from Black Light Trap
(Shriekback)

In the ballet of black light
And the agony of the fly

IN THE BLACK LIGHT TRAP
We make all our mistakes
IN THE BLACK LIGHT TRAP
Go in and out like tiny snakes

IN THE BLACK LIGHT TRAP
We fool around but that's okay
IN THE BLACK LIGHT TRAP
In the black light trap

Excerpts from Moths
(Jethro Tull, Heavy Horses)

The leaded window opened to move the dancing
candle flame
And the first Moths of summer suicidal came
And a new breeze chattered in its May-bud
tenderness

Creatures of the candle on a night-light-ride
Dipping and weaving—flutter through the
golden needle's eye in our haystack madness.
Butterfly-stroking on a Spring-tide high.

Life's too long (as the lemming said) as the candle
burned and the Moths were wed.
And we'll all burn together as the wick grows higher
before the candle's dead.

Insects in Literature

To Err Is Poetic

Insects are everywhere—even in poetry! It's probably not too surprising that animals as common as insects might have attracted the attention of a poet or two down through the ages.

Poets use insects in a variety of ways. In a line from "Haunted House," poet Thomas Hood created a sense of the supernatural: "And on the wall, as chilly as a tomb, The death's-head moth was clinging."

Frequently insects are used to make a philosophical comment or to teach a moral lesson. Such an approach was used by Solomon in his admonition: "Go to the ant, thou sluggard; consider her ways, and be wise." Shelly, in "Adonais," pens these worm-eaten verses:

We decay
Like corpses in a charnel; fear and grief
Convulse us and consume us day by day
And cold hopes swarm like worms within our
living clay.

But in spite of their assumed wisdom, poets frequently make mistakes about the insects that are the target of their vaunted pens. For instance, many poets ascribe the wrong sex to the insects in their writings. This is especially true of bees. Ingelow, in "Parnell," refers to the industry of the honey bee: "From sun to sun, from bank to bank he flies, With honey loads his bag, with wax his thighs." Of course, we all recognize that it is the female honey bee that does the work!

Even the Bard himself, William Shakespeare, made an entomological error when he assumed the

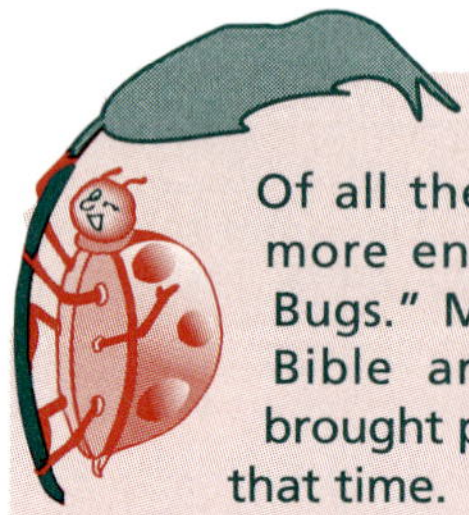

Biblical Curses

Of all the curses in the Bible, none were more enduring than the "Battle of the Bugs." Most of the 120 references in the Bible are of pest insects—those who brought plagues and curses to the people of that time.

The locusts were the leaders of the insect antagonists. These insects are a type of migratory grasshopper and are mentioned 24 times in the Bible. Plagues of locusts, even today, can be devastating in parts of Africa. In Exodus 10:15 read: "For they (the locust) covered the face of the whole earth, so that the land was darkened; and they did eat every herb of the land, and all the fruit of the trees . . ."

Locust devastation was predicted in Deuteronomy 28:38: "Thou shalt carry much seed out into the field, and shalt gather but little in; for the locust shall consume it." And "All thy trees and fruit of thy land shall the locust consume" (v. 42).

Flies also pestered mankind during biblical times. Flies are mentioned nine times in the Bible. Of the ten plagues of Egypt, six were insects and two were flies. Not just flies but "there came a grievous swarm of flies into the house of Pharaoh, and into his servants' houses," (Exodus 8:24).

And then there were lice. "And the Lord said unto Moses, Say unto Aaron, Stretch out thy rod, and smite the dust of the land, that it may become lice throughout all the land of Egypt" (Exodus 8:16).

Moths' damage was frequently used as an analogy. In Isaiah 51:8, we read the following: "For the moth shall eat them up like a garment, and the worm shall eat them like wool; . . . lo, they all shall wax old as a garment; the moth shall eat them up" (Isaiah 50:9).

Of course nothing could be more loathsome than to be devoured by worms. Thus in Job 21:26, "They shall lie down alike in the dust, and the worms shall cover them."

By modern standards, living in biblical times would not have been much fun, especially if locusts, moths and worms bug you.

head of the honey bee colony was a male. In "Henry V" we read: "They have a king, and officers of sort." And:

To the tent-royal of their emperor:
Who, busied, in his majesty, surveys
The singing masons building roofs of gold;

In this case, the head man turns out to be a woman. On the other hand, some poets made female insects out of males. Oliver Wendell Holmes wrote:

Thou art a female, Katydid!
I know it by the trill
That quivers through thy piercing note
So petulant and shrill.

Holmes failed to recognize—or at least admit—that it is only the male katydid that sings!

Thirty years before scientists confirmed it, Longfellow aluded to the relationship between mosquitoes and malaria. In "Hiawatha," Nakomis urges Hiawatha to:

Slay this merciless magician,
Save the people from the fever
That he breathes across the fen-lands
And avenge my father's murder!

Longfellow was partially correct. The merciless magician was the mosquito; however, in true poetic fashion, Longfellow blames the male mosquito for transmitting the disease. It is the female that bites and, thus, carries malaria.

We, of course, are obliged to excuse the entomological errors of poets—just call it poetic license!

Poems

References to insects are introduced into poetry for a variety of purposes, some of which are as follows:

As a figure of speech.

Here in her hairs
The painter plays the spider; and hath woven
A golden mesh to entrap the hearts of men,
Faster than gnats in cobwebs.

Shakespeare—*The Merchant of Venice*

Biblical Insects

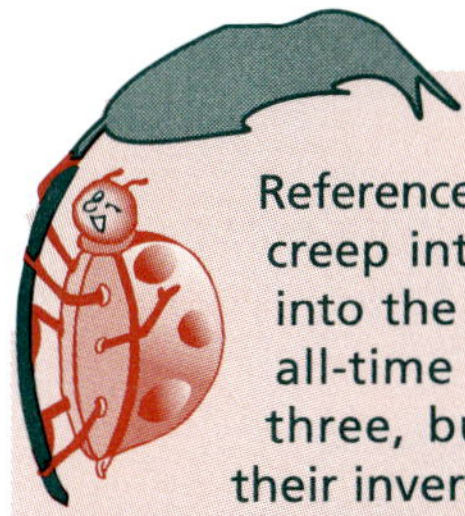

References to insects, like insects themselves, creep into the most unlikely places—even into the hallowed pages of the Bible. The all-time bestseller contains not two, not three, but 120 references to insects and their invertebrate relatives.

Most biblical references chronicle the negative side of insects—damage to crops, possessions and people. But in a few instances, the Bible speaks of insects in positive ways.

King Solomon used examples from the insect world to enhance his teaching. To the lazy idlers among us, Solomon recommends in Proverbs 6:6, "Go to the ant, thou sluggard; consider her ways, and be wise:" And what does wise, old Solomon expect us to learn from the ant's ways? In verses 7 and 8, Solomon points out that even though she has "no guide, overseer, or ruler, provideth her meat in the summer, and gathereth her food in the harvest."

In Proverbs, Solomon states that there are four things which are little upon the earth, but they are exceedingly wise. These small but wise things included: "The ants are a people not strong, yet they prepare their meat in the summer;" (30:25) and "The locusts have no king, yet go they forth all of them by bands" (30:27).

Some insects were used as food in biblical times. In Leviticus 11:22, we learn that approved food included, "Even these of them ye may eat; the locust after his kind, the bald locust after his kind, and the beetle after his kind, and the grasshopper after his kind." John the Baptist dined upon locusts and wild honey when he was wandering in the wilderness. Both Matthew (3:4) and Mark (1:6) recorded John's culinary habits in their Gospels.

It has even been suggested by some entomologists that manna—the heaven-sent food of biblical times—was actually produced by scale insects that grow abundantly in the region of the world in which biblical history was recorded. With or without "manna from heaven," people in biblical times did occasionally find insects to be useful. But to these ancient folks, insects were mostly troublesome, but that's another story.

In pastoral scenes as a part of rural life.

Now fades the glimmering landscape on the sight,
And all the air a solemn stillness holds,
Save where the beetle wheels his droning flight,
And drowsy tinklings lull the distant folds.

Gray—*Elegy*

To impress with a sense of the supernatural or weird.

The air was thick, and in the upper room
The bat—or something in its shape—
was winging;
And on the wall, as chilly as a tomb
The death's-head moth was clinging.

Hood—*Haunted House*

To use insect behavior as a subject for philosophic comment.

Dreams, empty dreams. The millions flit as gay
As if created only like a fly,
That spreads it motley wings in th' edge of
noon,
To sport their season and be seen no more.

Cowper—*The Task*

To accentuate the jest.

Your mouth, it was then quite a bait for the
bees,
Such nectar there hung on each lip;
Though now it has taken the lemon-like squeeze,
Not a blue-bottle comes for a sip!

Hood—*Hymeneal Retrospections*

That pest of the gardens, the little Turk,
Who signs with the crescent his wicked work
And causes the half-grown fruit to fall.

Bryant

Or as a swarm of flies in vintage time,
About the wine press where sweet must is
poured,
Beat off, returns with humming sound.

Milton—*Paradise Regained (Book IV)*

The housefly, stealing from his narrow grave,
Drugged with the opiate that November gave,
Beats with faint wing against the sunny pane,
Or crawls, tenacious, o'er its lucid plain.

Holmes—*Spring*

He, the mightiest of magicians,
Sends the fever from the marshes,
Sends disease and death among us!

Slay this merciless magician,
Save the people from the fever
That he breathes across the fen-lands
And avenge my father's murder!

All the air was white with moonlight,
All the water black with shadow,
All around him the Suggerma
The mosquito, sang his war song.

Longfellow—*Hiawatha*

Of the black-wasp's cunning way,
Mason of his walls of clay,
And the architectural plans
Of gray hornet artisans.

Whittier—*Barefoot Boy*

Let earth withold her goodly root,
Let mildew blight the rye,
Give to the worm the orchard's fruit,
The wheatfield to the fly.

Whittier—*The Corn Song*

The armyworm and the Hessian fly
And the dreaded canker-worm shall die,
And the thrip and the slug and the fruit-moth
seek
In vain to escape that busy beak,
And fairer harvests shall crown the year,
For the Old-World sparrow at last is here.

Bryant

Alas for the sparrow's vaulted "pep";
He has fallen down and lost his "rep";
And the armyworm and the Hessian fly
Ne'er cease to gnaw as he flutters by;
And tho' fair harvest garnered be
This is not due to such as he;
But science with her helping hand
Has put to flight this robber band.

Ade

In her suit of green arrayed,
Hear her singing in the shade—
Caty-did, Caty-did, Caty-did!

Freneau—*To a Caty-did*

This lady-fly I take from off the grass,
Whose spotted back might scarlet red surpass,
Fly, lady-bird, north, south, east or west,
Fly where the man is found that I love best.

Gay—*The Shepherd's Week*

Little inmate, full of mirth,
Chirping on my kitchen hearth,
Whereso'er be thine abode,
Always harbinger of good,
Pay me for thy warm retreat
With a song more soft and sweet;
In return thou shalt receive
Such as strain as I can give.

Bourne—*The Cricket*

How comes he to dis colda clime
To seeng so far from homa?
I catch him manny, manny time
W'en I am boy in Roma.
I catch heem een da fields an' tak'
Heem back eento da ceety,
Where reecha people try to mak'
Deir gardens fine an' pritty.
Dey are so glad for hear heem seeng
Dey no can gat too manny
An' so for evra wan I breeng
Dey geeva me a penny.
Dough here hees song ces justa same,
Hees name I no can speak eet—
Eh? w'at you call hees Anglaice name?
Ah! 'creecket,' yes, 'da creecket.'
Sh! nevva mind da snow,
An' how da weend ees blow:
Hoo-woo! Hoo-woo! Hoo-wee!
For here eet's warm, an 'O!
Il grillo seenga so:
Cher-ree! cher-ree! cher-ree!

Daly—*Il Grillo*

. . . Therefore doth heaven divide
The state of man in divers functions,
Setting endeavour in continual motion;
To which is fixed, as an aim or butt,
Obedience: for so work the honey bees;
Creatures that, by a rule in nature, teach
The act of order to a peopled kingdom.
They have a king, and officers of sort:
Where some, like magistrates, correct at home;
Others, like merchants, venture trade abroad;
Others, like soldiers, armed in their stings,
Make boot upon the summer's velvet buds;
Which pillage they with merry march bring home
To the tent-royal of their emperor:
Who, busied, in his majesty, surveys
The singing masons building roofs of gold;
The civil citizens kneading up the honey;
The poor mechanic porters crowding in
Their heavy burdens at his narrow gate;
The sad-ey'd justice, with his surly hum,
Delivering o'er to executors pale
The lazy yawning drone.

Shakespeare—*Henry V*

The shrilling locust slowly sheathes
His dagger voice, and creeps away
Beneath the brooding leaves where breathes
The zephyr of the dying day:
One naked star has waded through
The purple shallows of the night,
And faltering as falls the dew
It drips its misty light.
O'er garden blooms,
On tides of musk,
The beetle booms adown the glooms
And bumps along the dusk.

The katydid is rasping at
The silence from the tangled broom:
On drunken wings the flitting bat
Goes staggering athwart the gloom:
The toadstool bulges through the weeds;
And lavishly to left and right
The fireflies, like golden seeds,
Are sown about the night.
O'er slumb'rous blooms,
On floods of musk,
The beetle blooms adown the glooms
And bumps along the dusk.

Riley—*Dusk Song—The Beetle*

Blue dragon-flies knitting
To and fro in the sun,
With sidelong jerk flitting
Sink down on the rushes,
And, motionless sitting,

With level wings swinging
On green tasseled rushes,
To dream in the sun.

Lowell—*The Dragon-flies and Damsel-flies*

Little brook, sing to me:
Sing about a bumblebee
That tumbled from a lily-bell, and grumbled mumblingly
Because he wet the film
Of his wings, and had to swim
While the water-bugs raced round and laughed at him!

Little brook, sing a song
Of a leaf that sailed along
Down the golden-braided center of our current swift and strong,
And a dragon-fly that lit
On the tilting rim of it,
And rode away and wasn't scared a bit.

Riley—*The Brook Song*

Under the high-top sweeting,
Many a playmate came to share
The sports of our merry meeting:
Zigzag butterflies, many a pair,
Doubled and danced in the sunny air;
The yellow wasp was a visitor there;
The cricket chirped from his grassy lair;
Even the squirrel would sometimes dare
Look down upon us, with curious stare;
The bees plied fearless their honeyed care
Almost beside us, nor seemed aware
Of human presence; and when the glare
Of day was done, and the eye was fair,
The fireflies glimmered everywhere,
Like diamond-sparkles in beauty's hair,
In the boughs of the high-top sweeting.
The humming-bird, with his gem-bright eye,
Paused there to sip the clover,
Or whizzed like a rifle bullet by;
The katydid, with its rasping cry,
Made forever the same reply,
Which laughing voices would still deny;
And the beautiful four-winged dragon-fly
Darted amoung us, now low, now high,
And we sprang aside with a startled cry,
Fearing the fancied savagery
Of the harmless and playful rover.
The flying grasshopper clacked his wings,
Like castanets gayly beating;
The toad hopped by us, with jolting springs;
The yellow spider that spins and swings
Swayed on its ladder of silken strings;
The shy cicada, whose noon-voice rings
So piercing shrill that it almost stings
The sense of hearing, and all the things
Which the fervid northern summer brings—
The world that buzzes and crawls and sings—
Were friends of the high-top sweeting.

Akers—*The High-Top Sweeting*

'Tis a woodland enchanted!
The great August moonlight,
Through myriad rifts slanted,
Leaf and bole thickly sprinkles
With flickering gold;
There, in warm August gloaming,
With quick, silent brightenings,
From meadow-lands roaming,
The firefly twinkles
His fitful heat-lightnings.

Lowell—*The Fountain of Youth*

The shrill Cicadas, people of the pine
Making their summer lives one ceaseless song.

Byron

Here, in the summer, at a broken pane,
The yellow wasps come in, and buzz and build
Among the rafters; wind and snow and rain
All enter, as the seasons are fulfilled.

Here where the gray incessant spiders spin,
Shrouding from view the sunny world outside,
A golden bumblebee has blundered in
And lost the way to liberty, and died.

Akers—*In a Garrett*

The poetry of earth is never dead:
When all the birds are faint with the hot sun,
And hide in cooling trees, a voice will run
From hedge to hedge about the new-mown mead:
That is the grasshopper's he take the lead
In summer luxury—she has never done
With his delights; for, when tired out with fun,
He rests at ease beneath some pleasant weed.
The poetry of earth is ceasing never:
On a lone winter evening, when the frost
Has wrought a silence, from the stove there shrills
The cricket's song, in warmth increasing ever,
And seems to one in drowsiness half lost,
The grasshopper's among some grassy hills.

Keats—*The Grasshopper and the Cricket*

Ghosts of departed winged things,
What memories are those
That tempt you with your damask wings
Here where my candle glows?

Vainly you hover, circling oft
The tongue of yellow flame:
A tiger by caresses soft
You vainly seek to tame.

Here is no hope for you: nay, here
Death lurks within the light
To leap upon you flying near
And sweep you from the night.

Moon-butterflies, back to your blooms
Born of the dew and stars!
Hence, ghosts, and find again your glooms
Hidden by shadow-bars.

Quick—speed across the dusky blue,
Lest, in a sudden breath,
This tawny tiger wake, and you
Endure a second death!

Sherman—*Moths*

O the South Wind and the Sun!
How each loved the other one—
Full of fancy—full of folly—
Full of jollity and fun!
How they romped and ran about,
Like two boys when school is out,
With glowing face, and lisping lip,
Low laugh, and lifted shout!

Over meadow-lands they tripped,
Where the dandelions dipped
In crimson foam of clover-bloom,
And dripped and dripped and dripped;
And they clinched the bumble-stings,
Gauming honey on their wings,
And bundling them in lily-bells,
With maudlin murmurings.

And the humming-bird, that hung
Like a jewel up among
The tilted honeysuckle-horns,
They mesmerized, and swung
In the palpitating air,
Drowsed with odors strange and rare,
And, with whispered laughter, slipped away
And left him hanging there.

And the golden-banded bees,
Drowning o'er the flowery leas,
They bridled, reigned, and rode away
Across the fragrant breeze,
Till in hollow oak and elm
They had groomed and stabled them
In waxen stalls that oozed with dews
Of rose and lily-stem.

Where the dusty highway leads,
High above the wayside weeds
They sowed the air with butterflies
Like blooming flower-seeds,
Till the dull grasshopper sprung
Half a man's height up, and hung
Tranced in the heat, with whirring wings,
And sung and sung and sung!

And they loitered, hand in hand,
Where they snipe along the sand
Of the river ran to meet them
As the ripple meets the land,
Till the dragon-fly, in light
Gauzy armor, burnished bright,
Came tiling down the waters
In a wild, bewildering flight.

Riley—*The South Wind and the Sun*

I heard a fly buzz when I died;
The stillness round my form
Was like the stillness in the air
Between the heaves of storm.

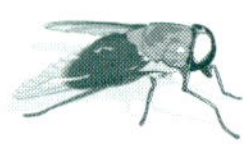

With blue, uncertain, stumbling buzz,
Between the light and me;
And then the windows failed, and then
I could not see to see.

Dickinson—*I Heard a Fly Buzz When I Died*

As the moths around a taper,
As the bees around a rose,
As the gnats around a vapor,
So the spirits group and close
Round about a holy childhood,
As if drinking its repose.

Browning—*A Child Asleep*

The firefly is a funny bug,
He hasn't any mind.
He blunders all the way through life
With his headlight on behind.

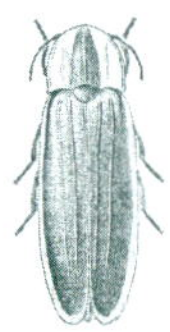

Prelutsky—
Poems of A. Nonny Mouse

There wouldn't be apples
on apple trees,
or daisies or clover
or such as these,
if it weren't for fuzzy old
buzzy old bees.

Fisher—*Buzzy Old Bees*

"What about caterpillars?
Where do they crawl"
when the stars say, "Frost,"
and the trees say, "Fall?"

Some go to sleep
in a white silk case
when the winds say, "Blow!"
and the clouds say, "Race!"

Fisher—*About Caterpillars*

Insects in Folklore

1. ***Zuni Indian Saying:***
"When the white butterfly comes, comes also the summer."
2. ***Proverb:***
"He who would gather honey must bear the sting of the bees."
3. ***Old American Saying:***
"If you kill a glowworm you will put the light out in your house."
4. ***Proverb:***
"The bee from his industry in the summer eats honey all the winter."
5. ***American Saying:***
"When the gnats swarm, rain and warmer weather are believed to be coming."
6. ***Proverb:***
"None preaches better than the ant and she says nothing."
7. ***Zuni Indian Saying:***
"When the white butterfly flies from the Southwest, expect rain."
8. ***Proverb:***
"What is good for the swarm is not good for the bee."
9. ***Trinidad Proverb:***
"Ants take over all de grease."
10. ***Blackfoot Indian Belief:***
"Dreams are brought to us in sleep by the butterfly (ap-u-nni)."
11. ***American Folk Belief:***
"It is a folk belief of long standing that local winter weather may be predicted by observing the width of the reddish-brown band of woolly bear caterpillars. The theory is that the narrower the reddish-brown band, the colder and longer will be the winter; the wider the band the milder the winter."
12. ***Ancient Proverb:***
"Patience and the mulberry leaf become a silk gown."
13. ***Proverb:***
"Stirring up a hornets' nest."
14. ***American Slang Expression:***
"A buggy person."

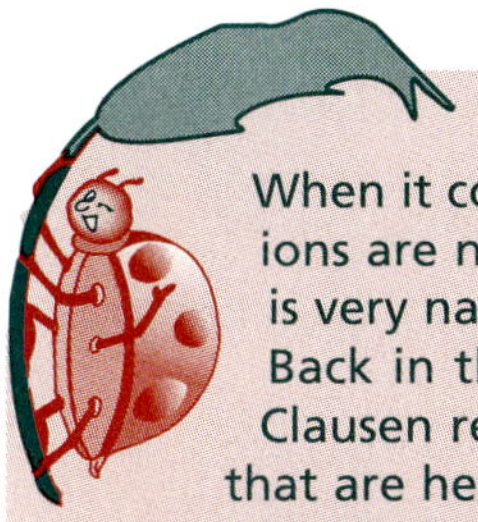

Myths About Insects Thick as Fleas on Dogs

When it comes to insects, one thing is certain. Everyone has an opinion! However, many of those opinions are not necessarily based on facts. When dealing with insects, the line between fact and folklore is very narrow. So narrow that it could reasonably be described as "thin as a gnat's eyelash!"

Back in the early 1950s, Lucy Clausen wrote a book called "Insect Fact and Folklore." In the book, Clausen recounts many tales about insects. For instance, the East Coast tale of how to kill mosquitoes that are helping themselves to a meal of blood. Our instinct is to swat those mosquitoes. But if you are an individual who has an abundance of self-control, try holding your breath. This is supposed to lock the mouthparts of the mosquito into your skin. You can then capture the insect at your leisure!

Another version of this mosquito-capture technique is to tighten your muscles around the proboscsis of the insect. The trapped insect can now be unceremoniously squashed. Or you can continue to hold the insect while more of your blood is shared with the offending beast. Rumor holds that eventually the mosquito will explode!

As you might imagine, the widespread presence and high numbers of cockroaches provide fodder for several urban legends. Two old beliefs, no doubt, fuel some of the modern myths about cockroaches. First, it is said that cockroaches will gnaw a person's toenails at night, unless the feet are covered with bedclothes. Another European legend holds that if a black beetle—another name sometimes used for cockroaches—flies against you, severe illness or even death will follow.

Such ideas probably lend support to the oft-repeated myth that someone licked the glue on an envelope, cut a lip and got cockroach eggs in the cut. The eggs then matured and eventually the individual had a boil develop at the site of the cut. Ultimately, the boil burst, and all kinds of little cockroaches emerged into the helpless individual's mouth.

Another version relates that the infestation originated from food purchased at some fast-food establishment. The general theme is the same. An individual gets a cut in the mouth while consuming the food. There are cockroach eggs on or in the food. And, well, you know the rest of the story.

Of course, there is absolutely no truth to any such story about cockroach eggs hatching inside a wound on a human. That is not the way cockroaches live. But such urban myths continue to exist.

Cockroaches aren't the only insects that have some rather gruesome lore associated with them. The insect commonly known as an earwig is another example. An old myth holds that this insect will crawl into a person's ear and bore through the eardrum. But it doesn't stop there—it will continue, actually eating into the human brain. Following which, the person goes mad! That myth is the basis for the common name of earwig.

Even fireflies do not escape negative folklore. One such myth holds that if the "fire" from a firefly gets into the eye, the eye will be put out. While the chemicals that a firefly uses to produce light taste bad, there is no evidence that they would cause such a thing to happen. Another saying holds that when a firefly comes into the house, on the following day, there will be one more or one less person in the house. While it is not clear what one more or less person really means, if it is a death or visiting relatives, it could be bad. But don't worry, it's just another insect myth!

---

15. Proverb:
"The fatter the flea the leaner the dog."

16. Proverb:
"Take not a musket to kill a fly."

17. Descriptive Language:
"Nit Picking!"

18. Proverb:
"Busy as a bee."

19. American Ditty:
"A swarm of bees in May is worth a load of hay. A swarm of bees in June is worth a silver spoon. A swarm of bees in July is not worth a fly (or just let them go by)."

20. Cherokee Indian Saying:
"Katydid has brought the roasting ear-bread."

21. Proverb:
"The diligence of the hive produces the wealth of honey."

22. Ancient Proverb:
"A worm is in the bud of youth and at the root of old age."

23. Irish Saying:
"Butterflies are souls of the dead waiting to pass through Purgatory."

24. Ancient Proverb:
"Cockroaches never get justice when the chicken is the judge."

25. ***Ancient Proverb:***
"Make yourself honey and the flies will devour."

26. ***American Superstition:***
"If one hive of bees is sold the contents of two hives will die."

27. ***American Saying:***
"Ants in your pants."

28. ***Midwest Saying:***
"When flies collect on the screen door it is going to rain."

29. ***Indiana Saying:***
"When the workers toss the drones out of the hive expect a long dry spell."

30. ***Midwest Saying:***
"Seeing caterpillars late in the fall predicts a mild winter."

31. ***Ancient Proverb:***
"The butterfly that brushes against thorns will tear its wings."

32. ***American Saying:***
"When swallows fly low, rain is on the way."

33. ***Ancient Saying:***
"A fly in the ointment."

34. ***American Folk Saying:***
"When hornets build nests near the ground a harsh winter is expected."

The Kite and the Butterfly

Ivan Andreyevich Krilov

A Paper Kite, which some boys were flying until it soared above the clouds, called down from on high to a Butterfly far below in the valley.

"Really and truly, friend Butterfly, I hardly recognized you from way up here. Confess, now, that you envy me when you see me flying."

"Envy you? No, indeed!" replied the Butterfly, "you have no reason for feeling so proud of yourself! You fly high, to be sure. But you are always tied to a string. Such a life my friend, is far from a happy one. As for me, humble though I am, I still can fly where I choose. I should not want to spend all my life as the tool of someone else's foolish amusement!"

Insects in the Fables of Aesop

Some 2,550 years ago in ancient Greece lived a fellow who was quite the storyteller. At least you would guess that Aesop probably wove a decent yarn.

His tales were good enough to have survived to this day and are often repeated. Indeed, there are few people who are not familiar with Aesop's Fables. The Fables of Aesop involve animals of all sorts. These creatures talk and act like humans, especially when it comes to doing really dumb things. And that is where the moral comes in!

Occasionally, insects creep into Aesop's Fables. The best known of these involves a grasshopper and an ant. You know the story. The grasshopper sang all summer while the ant worked hard putting up food for the winter. When winter came, the cold and hungry grasshopper showed up on the ant's doorstep begging for a handout. Of course, the ant said get lost. And Aesop's point was? Plan ahead or you've got to work for what you get.

Some people believe that it wasn't a grasshopper in the original fable. You see, grasshoppers are not the best known singers of the insect world. A better known singer, certainly in ancient Greece, was the cicada. So speculation is that it was the cicada, not the grasshopper, that according to Aesop was begging from the ant.

Ants are the most common insect in Aesop's Fables. There is "The Dove and the Ant," where the dove saves the ant from drowning in a stream. Later, the ant stings a hunter attempting to shoot the dove. The moral is "Little friends may prove great friends." Or in another version, "One good turn deserves another."

The ant shows up in another fable—this time as the bad guy! In "The Ant and the Chrysalis," the ant pities the poor chrysalis that is only able to wiggle its tail. The ant, on the other hand, can crawl up on the tallest objects. So the chrysalis turns into a butterfly and flies away, telling the ant to look at his most-pitied friend. Moral: "Appearances are deceptive."

In the fable of "The Bald Man and the Fly," a fly bit the bare head of the bald man who attempted to kill the fly with a slap. Of course, the bald-headed man missed and slapped his head in the process. The fly mocked the man, saying something like, "Look, you have hurt yourself trying to do me harm!" To which the man replies that he is willing to endure some pain in order to inflict even heavier damage on the fly. The lesson to be learned is that revenge will hurt the avenger.

Another fable attributed to Aesop is called "The Bee and Jupiter." In this fable, a queen from the hive ascended to Olympus to present Jupiter some fresh honey from her combs. Jupiter promised to give her what she wished for in return, and she asked for a sting to kill mortals who threatened her hive. This bothered Jupiter, but he granted the wish with the stipulation that the bee would die if she stung because the stinger would remain in the wound of the victim. The moral of the story is that "Evil wishes, like chickens, come home to roost!" In other words, when you do bad things there is always a penalty to pay.

Of course, bee experts will point out that queen honey bees do not die when they sting, but workers do. But let's not be too critical of old Aesop; at least he knew that some bees lose their stingers when they sting!

"The Lion and the Mosquitoes" (Chinese Folk Tale)

Once in the summer time the lion was very thirsty. But the sun had taken all the water near the lion's home, and he went to many places seeking it. In time he found an old well, but the water was not fresh. As the lion was very thirsty he said, "I must drink, even though the water is stale."

But when he reached down into the old well, he found that it was the home of all the mosquitoes of the wilderness.

The mosquitoes said to the lion, "Go away, we do not want you. This is our home and we are happy. We do not wish the lion, the fox, or the bear to come here. You are not our friend. Why do you come?"

The lion roared and said, "Weak and foolish things! I am the lion. It is you that should go away, for I have come to drink. This is my wilderness, and I am king. Do you know, weak things, that when I come out from my place and send forth my voice, all the creatures of the wilderness shake like leaves and bow their heads to me? What are you that you should have a place you call your home and tell me that I may or I may not?"

Then the mosquitoes answered, "You are only one. You speak as if you were many. Our people had this old well for a home before your roar was heard in the wilderness. And many generations of us have been born here. This home is ours, and we are they that say who shall come or go. And yet you come and tell us to go out of our own door. If you do not leave us, we will call our people, and you shall know trouble."

But the lion held his head high with pride and anger and said, "What are, oh, small of the small? I will kill every one of your useless people. When I drink, I will open my mouth only a little wider, and you shall be swallowed like the water. And tomorrow, I shall forget that I drank today."

"Boastful one," said the mosquitoes, "we do not believe that you have the power to destroy all our people. If you wish battle, we shall see. We know your name is great and that all animals bow their heads before you; but our people can kill you."

The lion jumped high in his rage and said, "No other creature in the wilderness has dared to say these things to me—the king. Have I come to the

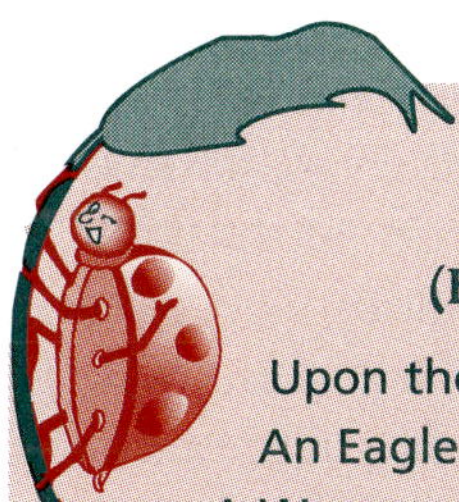

The Eagle and the Worm

(Russian) Anonymous

Upon the summit of a lofty rock,
An Eagle chanced to espy
A Worm; whom thus he gave in taunting tone to mock:
"Reptile! What raised thee thus high?
How haps it I so vile a creature see
Arched on the same imminence with me, Here daring to abide?"
"By my own strength," the Worm replied,
"I hither made my way; and small in
My opinion, the difference of the mode
In which to the same point we took our road,
What you by soaring did, I did by crawling."

Insect Weather Predictors

Everybody is interested in the weather. If we aren't talking about it, we are trying to predict it. Today, weather forecasters use a variety of scientific tools and equipment, but that hasn't always been the case. There was a time when nature provided the primary information used in weather predictions. For instance, insects and insect behavior.

Flies, it is said, bite sore just prior to a rain. When rain is on the way, gnats swarm and flies collect on the screen door. Swallows also fly low just before a rain or a change in the weather. This is because the swallows are feeding on flying insects that also fly low just before a rain. Butterflies flying from the southwest also indicate the approach of rain.

Folklore holds that stepping on an ant will bring rain. Maybe that ancient belief gave rise to the Indian rain dance when some overzealous individual got into an ant nest!

Insects have been used to predict the onset of the seasons. It is said that a yellow butterfly flying in one's face indicates a frost within the next 10 days that is sufficient enough to turn the leaves the color of the butterfly. The first song of the dog day cicada means six weeks to frost. A Zuni Indian saying indicates, "When the white butterfly comes, comes also the summer."

Where a butterfly chrysalis is suspended is said to indicate weather trends. If the chrysalis is on a heavy branch, expect rain. If it is found on the underside of slender branches, then we are in for a spell of fair weather.

Foraging behavior of bees gives a hint of weather to come. Someone put these ideas in a poem:

When bees to distance wing their flight,
Days are warm and skies are bright.
But when the flight ends near their home,
Stormy weather is sure to come.

(Hopefully the writer's observation of nature was better than his poetry.)

The insect really comes into its own in predicting the severity of winters. A harsh winter is sure to follow when bees lay up an unusually large store of honey. Hornets, it is said, build nests near the ground when a severe winter is expected.

The champion of the insects used to predict winter has to be the woolly-bear caterpillar. This fuzzy black caterpillar, with a reddish-brown band around its midsection, wanders around in the fall looking for a place to spend the winter. According to legend, the narrower the band, the longer the winter. However, don't sell the snowplow based on wide bands on woolly-bears; scientists say the width of the band has to do with the genes the caterpillar received from its parents!

---

vile well of the silly mosquitoes for wisdom?" And he held his head high, and gave the mighty roar of battle, and made ready to kill all the mosquitoes.

Then the mosquitoes, big and little, flew around him. Many went into his ears, and the smallest ones went into the nose, and the big old ones went into his mouth to sting. A thousand and a thousand hung in the air, just over his head, and made a great noise, and the lion soon knew that he could not conquer.

He roared and jumped, and two of his front feet went down into the well. The well was narrow and deep and he could not get out, for his two hind feet were in the air and his head hung downward. And as he died, he said to himself:

"My pride and anger have brought me this fate. Had I used gentle words, the mosquitoes might have given me water for my thirst. I was wise and strong in the wilderness, and even the greatest of the animals feared my power. But I fought with the mosquitoes and I die—not because I have not strength to overcome, but because of the foolishness of anger."

Ee-Sze (meaning): The wise can conquer the foolish. Power is nothing, strength is nothing. The wise, gentle, and careful can always win.

Insects in Cartoons and Children's Literature

Bugfolk

We human beings sometimes impart our characteristics to other animals, a habit called anthropomorphism. Through the ages, we have conjured up creatures like the Centaur, which was half man and half horse, and the Mermaid, a combination of woman and fish. We have even humanized insects, producing a creature that Charles L. Hogue of the Natural History Museum of Los Angeles County calls "Bugfolk."

There aren't, however, many bugfolk, because it's hard for humans to imagine themselves in something as lowly as an insect. Some of the early bugfolk in human history were merely humans with insect wings. One of the earliest was the Greek Goddess Psyche who represented the soul. She was depicted as possessing butterfly wings. Aesop's

The Insects in Children's Poems by James Whitcomb Riley

James Whitcomb Riley, the Hoosier Poet, is well known for his poems for children. His book "Joyful Poems for Children" was first published in 1892 and has been reprinted many times.

Riley's collection of poems for children include some of his best-known works. Some of the poems are about people. The Raggedy Man, Little Orphant Annie and Aunt Mary from "Out to Old Aunt Mary's" are personalities that most children recognize.

Many of Riley's poems are childhood memories: for instance, that wonderful poem about the fall season that begins "When the frost is on the punkin and the fodder's in the shock." Going fishing in "The Fishing Party," or swimming at "The Old Swimmin'-Hole" or just walking in a stream in "The Brook Song" are sweet memories of carefree days of our youth.

Riley also writes about animals. "The Pet Coon," "The Tree Toad," "The Yellow Bird" and "Old Bob White" recognize the importance of animals to children.

Ever a student of nature, Riley also sprinkles his verses with references to insects. He describes the sound of katydids as "cheep." In "Summer-time and Winter-time," he writes that in summer "Open eyes, or drowsy lids, 'Wake or 'most asleep, I can hear the katydids—Cheep! Cheep! Cheep!" In "The Katydids," Riley uses the sound of the katydids to put him to sleep.

In "The Old Swimmin'-Hole," dragonflies are called snake-feeders.

"And the snake-feeders four gauzy wings fluttered by
Like the ghost of a daisy dropped out of the sky,
Or a wownded apple-blossom in the breeze's controle
As it cut acrost some orchurd to'rds the old swimmin'-hole."

In "A Sudden Shower," the plight of the insects in the rain is considered.

"The swallow dips beneath the eaves
And flirts his plumes and folds his wings;
And under the Catawba leaves
The caterpillar curls and clings.
The bumble bee is pelted down
The wet stem of the hollyhock;
And suddenly, in spattered brown,
The cricket leaps the garden-walk."

In "Little Orphant Annie," Riley uses the lack of insect activity as a sign of danger in these lines.

"An' you hear the crickets quit, an' the moon is gray,
An' the lightnin'-bugs in dew is all squenched away,—"

Riley also learned about bumble bees in a poem of that name. He learned as the first line says: "You better not fool with a bumble bee!—Ef you don't think they can sting—you'll see!" Riley experienced the truth about bees by grabbing one. This resulted in a stinger stuck in his skin, which was removed by the beloved Raggedy Man who admonished, "I told ye so!"

No doubt James Whitcomb Riley was stung by a bee, but it wasn't a bumble bee. The truth is that a bumble bee does not leave its stinger. That happens when a honey bee stings. Another entomological error also occurs in that poem. With apologies to Mr. Riley the poet, "He also called the bee a he, but in truth a stinging bee's a she!"

fables included insects with the ability to use human language. In Shakespeare's "A Midsummer Night's Dream" elves were depicted with butterfly wings.

Yet Jiminy Cricket is the most recognizable of the bugfolk. Jiminy, that loveable little creature with top hat and umbrella, played a starring role in Disney Studio's film "Pinocchio." Except for his small size and his ability to sing, Jiminy exhibits few insect characteristics. However, viewers of the film have little doubt that Jiminy is indeed . . . a cricket.

Modern cartoonists have made good use of bugfolk. Hart uses ants frequently in his "B.C." comic strip. Hart's ants live in anthills, are sometimes zapped by anteaters and have insect-like antennae protruding from their heads. Instead of six legs, the insect complement, these creatures have two arms and legs. That's probably the way it should be since the ants in Hart's comic strip are beset with all kinds of human problems that they address with the full range of human emotions.

Gary Larson is the king of modern cartoonists when it comes to the use of bugfolk. Such creatures are commonplace in his "Farside" cartoon. Larson's insect characters cover a wide range of types and are anatomically correct with antennae, six legs,

The Butterfly's Ball and the Grasshopper's Feast

"Come, take up your hats, and away let us haste to the Butterfly's Ball and the Grasshopper's Feast: The trumpeter, Gadfly, has summon'd the crew, and the revels are now only waiting for you."

So said little Robert, and, pacing along, his merry companions came forth in a throng; and on the smooth grass, by the side of a wood, beneath a broad oak that for ages had stood, saw the children of earth and the tenants of air for an evening's amusement together repair.

And there came the beetle, so blind and so black, who carried the Emmert, his friend, on his back;

And there was the Gnat and the Dragonfly too, with all their relations, green, orange, and blue;

And there came the Moth, with his plumage of down, and the Hornet, in jacket of yellow and brown, who with him the Wasp, his companion, did bring, but they promised that evening to lay by their sting.

And the sly little Dormouse crept out of his hole, and brought to the feast his blind brother, the Mole.

And the Snail, with his horns peeping out of his shell, came from a great distance—the length of an ell.

A mushroom their table, and on it was laid a water dock leaf, which a tablecloth made.

The viands were various to each of their taste, and the Bee brought her honey to crown the repast.

Then close on his haunches, so solemn and wise, the frog from the corner look'd up to the skies; and the squirrel, well pleased such diversions to see, mounted high overhead, and look'd down from a tree.

Then out came the Spider, with finger so fine, to show his dexterity on the tight line. From one branch to another his cobwebs he slung, then quick as an arrow he darted along; but just in the middle—oh, shocking to tell!

From his rope in an instant poor harlequin fell; yet he touch'd not the ground, but with talons outspread hung suspended in air at the end of a thread.

Then the Grasshopper came with a jerk and a spring, very long his own praises the rest of the night;

He took but three leaps and was soon out of sight, then chirp'd his own praises the rest of the night.

With step to majestic the snail did advance, and promised the gazers a minuet to dance.

But they all laugh'd so loud that he pull'd in his head, and went in his own little chamber to bed.

Then, as evening gave way to the shadows of night, their watchman, the Glowworm, came out with a light.

"Then home let us hasten, while yet we can see, for no watchman is waiting for you and for me." So said little Robert, and, pacing along, his merry companions return'd in a throng.

William Roscoe Esq. (1806)
Parliament for Liverpool
Member of Gentlemen's Quarterly

two or four wings, mandibles and distinct body segments. However, Larson's insects are folk, because they walk upright and talk. His bugfolk frequently address issues pertinent to insect life such as shedding of the exoskeleton or food habits, but always with the moral issues that only we mere mortals can appreciate.

Bugfolk are showing up in increasing numbers today to help us see ourselves and even get us to consider some profound truths. Of course the real question of a truth learned from a bugfolk is whether the truth was from the insect or the human portion of the creature.

Shopping for Insect Stuff on the Net

Well, I did it. I bowed to peer pressure and personal curiosity and went shopping on the net. I really wanted to find out if any insect stuff was available. So, I checked out the net auction company, eBay. There were 29 listings under beetles, but only nine turned out to be of the insect variety. The other beetles included such things as a beetles doll of George Harrison. A beetles record. And a beetles fan button pin. Never mind that the British rock group spelled their name with an "a," as in Beatles! Under beetles I found some VW concept beetles. Also a real "bug" automobile.

In the antique market, butterflies overshadow beetles. There were 260 hits for butterflies. All kinds of butterfly things. Ceramic hand-painted butterflies. Cross-stitch books to make your own butterflies. Butterfly stamps, butterfly paper weights, bamboo butterfly coasters and butterfly-design pinochle playing cards.

There was also butterfly pattern U.S. glass offered. One example was called "Flower Garden with Butterflies." I've got a feeling that the flowers are more prominent than the butterflies. One seller mentioned that the fun part is finding the butterflies hidden among the garden. Maybe, but if I'm buying butterflies I want to see them, not have to look for them!

There were two lovely Lalique-signed butterflies. Offered appropriately by the "Butterfly Lady." There was also a motion graphics kind of deal where two monarch butterflies are flapping their wings while resting on "Thank You." Oh, yes, below the "Thank You," the butterfly lady flaps about. One is to presume it is not the same butterfly lady who was offering Lalique!

An Amphora Vase, called "Butterflies and Bees" and described as Art Nouveau, had a high bid of $848 but the minimum had not been met. Ever notice how items described in hard-to-pronounce words sell higher? There was also a piece of Herend china called "Charger Birds and Butterflies." Three birds and 13 butterflies to be exact. The price at the time was $396.

Listed under butterflies was a 1972 playbill for "Butterflies are Free" with Gloria Swanson. There were also salt and pepper shakers, old crochet pillowcases and hankies, all with the butterfly motif.

For the high-fashion set, a Gianni Versache scarf was offered. It was adorned with puppies and butterflies with a few ladybugs thrown in for good measure. How about some vintage topaz butterfly pins to adorn the Versache scarf? They were available, as was a cuff links and tie bar set.

Flies included 130 items. Mostly fishing flies. But also a "Lord of the Flies" book, an Alice in Chains—Jar of Flies CD and a 1972 movie, "Four Flies on Grey Velvet."

I thought I had made a real find when I listed ants. There were 383 items. Unfortunately, most were "Ants—the anteater" Beanie Babies! But there was also an Adam and the Ants LP album called "Ant Rap," an "Empire of the Ants" movie starring Joan Collins and even an "Ants in the Pants" game.

I didn't buy anything, not even any of the insect stuff. I guess you could say I'm just sort of an electronic window shopper!

Insects Infest Upscale Catalog Offerings

There are times when upscale catalogs are thicker in our mailbox than fleas on a skinny hound dog. Gone are the days when one catalog offered everything from underwear to automobiles. The thick tome of the days of Sears and Roebuck has been replaced.

Nineties' catalogs are specialty books—not just clothing but natural fiber clothing! Not furniture, but wicker furniture! You name it, and there's bound to be a catalog for it.

I don't order anything from these catalogs, but I do look through them, all the while wondering, "Who would buy such a thing?" I have happened to notice that there are a fair number of insect things creeping into these high-falutin' sales rags.

It is probably not surprising that the National Wildlife Federation would include some insect offerings in its "sharing nature" catalog. There was a ladybug lamp. Use it as a nightlight or accent lamp.

Butterflies are big with the wildlife folks. There was a butterfly garden pullover, with the butterflies identified. Also, an abalone shell butterfly watchband. Need candle holders? How about dragonfly wall sconces that provide elegant lighting with the candles held in place by the curled abdomen of the dragonflies?

The J. Jill slick-pager offered decorative cast masonry tiles featuring a moth, grasshopper or dragonfly. Bees, dragonflies and ladybugs showed up in lightweight hairpins. Then, there was the insect hook rack. Not a place to hook insects, but where insects shaped as hooks hold things like car keys and garden tools, while the owner worked in the garden. Handmade by Amish craftsmen according to the advertisement.

Garnet Hill offered butterfly bedding stuff. Sheets, pillowcases and comforter covers. Doesn't go with your bedroom decor? How about a green "bugs bed?" Those bed and butterfly covers would give real meaning to the saying about not letting the bed bugs bite. Need a rug for your insect decorating? Garnet Hill has the ticket. A butterflies round rug, hand-hooked in pure wool, ought to fill the bill.

Even the matron of mauve, Martha Stewart, has discovered insects. Not in her home as pests, we trust, but as items sold in her Martha by Mail catalog.

Martha says decorate your kid in a crepe-paper monarch butterfly—they were even featured in "Martha Stewart Living." Speaking of decorating, use wood dragonflies, bees and butterflies, all mounted atop wood rods with a spring so they flutter, to add a little something to your decor.

The really domestic among us will no doubt want to make soap with assorted butterflies embedded inside. The soap will go with the butterfly mobile that you also have to make. Hey, you can do it. It's for ages 7 and up! Want to bake instead? How about making some cookies using Martha's butterfly cookie cutters? You can set them on a platter right next to the beehive honey pot.

Nothing out of place at Martha's house, not even pictures. So, hang them up using a butterfly picture hanger. Martha saw it at an antiques fair, so now you can buy a reproduction, through her catalog, of course. Got a bulletin board? Keep things in place using butterfly pushpins.

Even the Martha by Mail registered trademark is insect related. It is a bee skep and a bee. Martha isn't the first person to adopt an insect as a symbol. Napoleon also had the bee as his symbol. But I don't believe he had a catalog!

FIGURE 22.1. *Butterfly Commemorative Stamp.*

Insects in Arts and Crafts

Artists from ancient to modern times have included images of insects in their work. Insects are nearly as pervasive in art and crafts as they are in nature.

Insects are common in jewelry. One of the most common insects used as a motif in jewelry is the scarab beetle. This sacred insect of the ancient Egyptians is a dung beetle, seemingly not the kind of insect most people would want as jewelry. Because of the religious significance of this insect the ancient Egyptians used them in burial rituals. From there the insect itself and carvings of the insect came to adorn living humans. Dung beetles—excuse me—scarab beetles are still popular today as jewelry.

The great age of entomological jewelry was the Victorian era. During the reign of Queen Victoria of England (1837 to 1901) folks had a great interest in nature and ornate design in, among other things, architecture, furniture and jewelry.

Buttons also have been adorned with insects. As in most art butterflies are the most common insect depicted in buttons. The next most numerous in-

Insect Ties Make Fashion Statements

I can think of no item of apparel as useless as a necktie. How dumb does a person have to be to loop a piece of cloth around their neck and tie it in a knot? Of course, we do lots of dumb things in the name of fashion, and wearing a tie is just one of them. There are, however, a few people who benefit from this apparently insane habit—tie designers and manufacturers and clothiers for example.

Ties also provide a source of gifts for those who think a tie is just what every man desires as a present on some special occasion. At least size is not a problem when a tie is the gift and, as gifts go, ties are easy to wrap.

These nonfunctional hunks of cloth do provide a wonderful opportunity for the wearer to make a statement. That statement might be dictated by the business world. For instance, red ties are considered power ties. Blue striped ties are traditional and broadcast the stability of the wearer. Tie motifs can represent schools or geographical regions. Some ties represent occupations, such as those displaying airplane or auto designs. Other motifs, like golf clubs or fishing lures, are suggestive of hobbies.

There are also creature ties. Almost any animal on the surface of the earth has shown up in a tie motif, including insects. All kinds of insects have graced ties over the years. I have several insect ties. Some were gifts; some I actually purchased myself. It seems that anyone who studies insects ought to have an insect tie or two.

Butterflies seem to be the most common insect to appear on ties. Some butterfly ties are quite gaudy, with big, bright-colored butterflies. Other butterfly ties are more sedate, with smaller butterflies gracing the tie in repeating patterns. In fact, you have to look closely to notice that the design is actually an insect. I have such a tie with a red background. A power tie? Maybe, but a butterfly would seem to project too fragile an image for the power monger.

I have a butterfly tie that was specially designed and given to me as a gift by friends. I also have an M.C. Escher tie that is a typical Escher design in that it features a repeating motif—in this case moths. I have the reddish version, but blue and purple versions also exist.

Some insect ties feature bees and wasps. These stingers should be on power ties. After all, even Napoleon recognized the bee's industry and sting, choosing it as the symbol of his reign. Mostly, bees seem to show up on blue ties. While I can appreciate the traditional aspect of the blue color, it would seem that a few bees on a red tie would really send a message.

Beetles also appear on ties, especially ladybugs. One of my ladybug ties features blue insects on a yellow background. I've never seen a blue ladybug in my life, but the tie is rather attractive.

There are ties with ants. I have a tie with black ants walking over a red-and-white-checked tablecloth pattern. I don't know what this says about me, but when I wear it even strangers are motivated to comment. They say things like "Nice tie!" or "Where did you get that tie?" I take those comments as compliments.

Almost all kinds of insects have been included in tie designs. It makes sense. If you have to wear a tie, you might as well wear one that really bugs you!

Beetle Brooches and Diamond Dragonflies

Many people have "worn" insects under protest, as the little creatures look for a resting place during their travels. Other people choose to wear insects . . . in the form of jewelry, that is. Our chief ecological competitors are used in almost all societies to create articles of adornment.

Many beetle species, with their brilliant colors, are a popular motif in jewelry, appearing as necklaces, pins, earrings and the like. But while most of these items are a mere facsimile, a few are the genuine article. In fact, the idea of wearing live insects as jewelry probably inspired the insect designs found in the jewelry made of precious metals and gems.

Live jewelry can still be found today. Some beetles are extremely hardy and do not feed in the adult stage. Such insects are commonly collected in Mexico and, with rhinestones and a delicate chain, cemented to a wingcover to become a living brooch. Tourists who purchase such a piece of jewelry seldom get to show it to the relatives back home though. Federal law prohibits importing live insects into this country without a permit. Consequently, live-beetle brooches are confiscated at border inspection stations.

Probably the first insects to be used as jewelry were the scarab beetles of ancient Egypt. These dung beetles were thought to have magical powers and were worn by soldiers into battle for protection. The scarab is still a common motif for jewelry. Indeed, many high-society ladies decked out in the finest of designer clothing have attended a gala affair wearing none other than a jeweled dung beetle.

Another popular beetle, the common lady beetle, has been used as a cover for a fine, Swiss, pendant watch. When the antennae of the insect were compressed, its wings spread to expose the face of the watch.

Of course, beetles are not the only insects to show up in jewelry. Butterflies, bees, and dragonflies are commonly depicted in jewelry. And termites and flies have dangled from human earlobes under the guise of jewelry.

Insect jewelry was common during the Victorian era, when there was a great, popular interest in insects and other products of nature. Some of the finest examples of Victorian insect jewelry were included in the sale of the late Dr. Moser Lyon Stadiem of New Orleans in 1969. Some of his brooches were described in the sale brochure as "bizarre rarities with wings that seemed to fly on their hidden coil springs at the slightest movement." Most items sold for a few thousand dollars, but one item—an 1890 diamond dragonfly brooch—brought $27,500.

The use of insects as jewelry would probably come as no surprise to many children who grew up in areas where cicadas were common. Many kids have attached to their clothing the newly emerged cicadas or their shells. In most people's minds, cicadas make a first class brooch . . . well, at least they would in comparison to dung beetles.

---

sects on buttons are beetles, followed by flies, bees, dragonflies and grasshoppers.

M. C. Escher even let insects creep into his renderings. His "Infinity" shows ants locked in an endless path around the infinity symbol. High atop historical old Faneuil Hall in downtown Boston is a weathervane that was modeled after a grasshopper. That grasshopper weathervane has been there since 1742.

Stamps frequently depict insects. In the U.S. one set of butterfly commemorative stamps was issued at Indianapolis, IN on June 6, 1977. Worldwide there are literally thousands of stamps with an insect motif.

Native Americans also admired, and sometimes worshiped insects. The Hopi of the Southwest are well known for the butterfly dance. This dance was a supplication for good crops. Other dances suggestive of arthropods include the tarantella which is said to be a lively folk dance that resembles a nervous affectation caused by the bite of the tarantula. A more modern version of such uncontrolled dancing is known as the jitterbug!

Glossary

abdomen—last major body division, posterior to thorax

anal area—posterior region of the wing that lies closest to the body

Anoplura—literally "unarmed tail," insect order commonly known as the sucking lice

antenna (pl. antennae)—sometimes called "feelers," paired sensory structures located on the head

anterior—the head end of the insect

aptera—wingless

bugfolk—the giving of human characteristics to insects

capitate—antennae with knobs on the end

cardiac glycosides—bitter chemical from the milkweed plant sometimes used as a heart medicine

carrying capacity—often represented by the letter "K" on graphs, the maximum number of individuals that can survive in a particular environment

caterpillar—from the Latin *catta pilosa*, which literally means "hairy cat," terms used to refer to the larval stages of Lepidoptera

cerci—projections protruding from the posterior end of the abdomen

chitin—nitrogenous polysaccharide (or protein) that provides structure to the exoskeleton

chrysalis—the pupal stage of the butterfly

Coleoptera—literally "sheath wings," insect order commonly known as beetles

Collembola—literally "glue bolt or wedge," this insect order is commonly known as springtails, some scientists do not classify Collembola with the insects, the order name is a reference to the collophore

collophore—structure on the ventral surface of the first abdominal segment that is involved in water uptake

compound eye—the multi-lensed eye of the insect, used for seeing objects

cornicle—tubular structure located on the posterior end of the abdomen, usually paired and found on aphids

crop—structure of the digestive system just before the stomach

density dependent factors—factors that can influence a population size based on the size of that population, such as predation, competition, or disease

density independent factors—factors separate from population size that may impact that population size, such as bad weather in the form of excessive heat or hurricanes

Dermaptera—literally "skin wings," insect order commonly known as earwigs

Dictyoptera—literally "net wings," insect order commonly known as cockroaches, praying mantids, and walking sticks

Diplura—literally "two tail," insect order known as diplurans, some do not classify Diplurans with the insects

Diptera—literally "two wings," insect order commonly known as the true flies

dorsal—the top surface of the insect

ecdysis—or molting, the process of shedding the exoskeleton

economic threshold—level where a pest population is controlled to prevent that population from exceeding the economic injury level, also called action threshold

elytra—hardened modified front wings of beetles

Embioptera—literally "lively wings," insect order commonly known as web spinners, some classify the Embioptera as Embiidina

entomology—the study of insects

entomophobia—an unfounded fear of insects

Ephemeroptera—literally "for a day wings," aquatic insect order commonly known as mayflies in which the adult may live from only several hours to a few days

equilibrium position—average population of an organism over time

ethyl acetate—the chemical used in killing jars to kill insects that are collected

exoskeleton—outer covering of the insect which provides structural support

exuviae—cast skin of the insect, sometimes called a shuck (especially by fly fishers)

filiform—thread-like antennae

forensic—belonging to courts of judicature or to public discussion or debate, most often associated with crime investigation

frass—insect poop

geniculate—antennae that are bent, or "elbowed" in the middle, most often seen in ants

glowworms—immature stages of the firefly

glycerol—chemical used by some insects as anti-freeze in their bodies to prevent freezing during the winter

gnats—general term for very small flies

grubs—beetle larva

haltere—modified hindwings used as balancing organs by the Diptera, usually with a broadly rounded tip

Hemiptera—literally "half wings," insect order commonly known as the true bugs and includes stink bugs, milkweed bugs, assasin bugs, and bedbugs

hexapod—six legged organisms, a term sometimes used for the insects

Homoptera—literally "same wings," insect order commonly known as the cicadas, leafhoppers, treehoppers, scale insects, and aphids

horn—term generally used when referring to the antennae, for example, the long-horned grasshoppers

Hymenoptera—literally "membrane wings," insect order commonly known as bees, wasps, and ants

insecticide—specific pesticide used to control insects

interspecific competition—competition among individuals of different species

intraspecific competition—competition among individuals of the same species

Isoptera—literally "equal wings," insect order commonly known as termites

labium—mouthpart structure considered the bottom "lip" of the insect

labrum—mouthpart structure considered the top "lip" of the insect

lamellate—antennae with "flags" or finger-like projections at the end, often seen in certain beetles

larva (pl. larvae)—term used when referring to the immature stages of insects

Lepidoptera—literally "scale wings," insect order commonly known as the butterflies and moths

logarithmic rate of growth—J-shaped curve that relates population growth to time

loopers—the larvae of the geometrid moth that forms a loop with its body when it walks

luciferin—protein involved in light production of the firefly

luciferase—enzyme that uses luciferin in the presence of oxygen that fireflies use to produce their light

maggots—term used when referring to the larval stage of the true flies (Diptera)

Mallophaga—literally "wool eaters," insect order commonly known as chewing lice

malpighian tubules—insect kidney

mandible—the chewing structures of the insects' mouth, analogous to human teeth

maxilla (pl. maxillae)—insect mouthpart structure usually with palp that is used to manipulate food items, much in the same way humans use their hands when eating

Mecoptera—literally "long wings," insect order commonly known as scorpionflies

Megaloptera—literally "large wings," aquatic insect order of large insects commonly known as dobsonflies and fishflies, some have classified Megaloptera with the Neuroptera

metamorphosis—"change in form," term used when referring to the development of an insect from the immature stage to the adult, and includes two types, incomplete and complete metamorphosis

molting—or ecdysis, the process of shedding the exoskeleton

naiad—term used by some when referring to the immature stages of aquatic insects with incomplete lifecycles, such as dragonflies and mayflies

Neuroptera—literally "nerve wings," insect order commonly known as lacewings and ant lions, some classify the Megaloptera with the Neuroptera

nymph—term used when referring to the immature stages of certain insect with incomplete metamorphosis, including the Orthoptera and Hemiptera

ocelli (singular ocellus)—or "eye spot," the single lens eye generally used for detecting light

Odonata—literally "tooth," aquatic insect order commonly known as dragonflies and damselflies, the order name is a reference to the teeth on the mandibles

Orthoptera—literally "straight wings," insect order commonly known as grasshoppers, crickets, and katydids

ovipositor—egg laying structure found in some female insects, especially in the Orthoptera and Hymenoptera

parthenogenesis—reproduction without mating

peritrophic membrane—when food enters the stomach it is encased in this sac

pheromone—chemical signals insects use to communicate with one another

phylogeny—the evolutionary history of an organism showing its relationships with other organisms

phylogenetics—the science of studying the evolutionary relationships between organisms

phytophagous—organisms that feed on plants

phytotoxins—noxious plant chemicals used for defense by plants

Plecoptera—literally "folded or plated wings," aquatic insect order commonly known as stoneflies

plumose—antennae that are fuzzy or feathery, often seen in male moths

posterior—the back end of the insect

proboscis—any extended mouthpart structure, commonly found in the Diptera, Hymenoptera, and Lepidoptera

prolegs—unsegmented, fleshy structures used for gripping or locomotion, generally found in caterpillars and some maggots

pronotum—first segment of the thorax

Protura—literally "first tail," insect order of minute insects known as proturans, some scientists do not classify Proturans with the insects

proventriculus—gizzard, grinding structure of the digestive system

Psocoptera—literally "rub small wings," insect order of small insects known as booklice and barklice, the order name initially was a reference to the chewing habits of these insects

pupae (singular pupa)—"resting stage" of insects with complete metamorphosis which occurs between the larval and adult stage. In this stage the insect completes development into the adult

pyrethrin—an extract from chrysanthemums now used in insecticides

saprophagous—organisms that feed on decaying organic matter, scavengers

scales—flattened, modified hairs usually located on the wings of the Lepidoptera, but may occur on the bodies or wings of certain other insect groups

Siphonaptera—literally "tube wingless," insect order commonly known as fleas, the order name is a reference to their mouthpart structure and wingless condition

spiracle—respiratory structures, holes in the cuticle that connect to the trachea

Strepisiptera—literally "twisted wings," insect order of very small insects known as the twisted-wing parasites

stridulation—rubbing one body part against another to produce a song or a chirp, commonly occurs in crickets and grasshoppers

style—usually singular projection extending from tip of abdomen

sylopids—Strepisiptera that are parasitic on certain bees and wasps

tanning—process of hardening the new exoskeleton after molting is complete

taxonomy—the science of describing and classifying organisms

taxonomist—a scientist who describes and classifies organisms based on their evolutionary relationships

tails—segmented appendages projecting from tip of abdomen

tarsi (sg. tarsus)—portion of the leg between the tibia and claw, usually made up of several segments

thorax—middle body region located between the head and abdomen and where the legs and wings are attached

thrips—insects of the order Thysonoptera

Thysanoptera—literally "fringe wings," insect order of small insects commonly known as thrips

Thysanura—literally "bristle tail," insect order commonly known as silverfish and firebrats

trachea—series of tubes in the insect used for breathing, connected to the outside via spiracles

Trichoptera—literally "hair wings," aquatic insect order commonly known as caddisflies, these insects are closely related to the Lepidoptera

tumbler—mosquito pupa, so called because of the "tumbling" action it makes as it moves through the water

ventral—the bottom side of the insect

wanderers—caterpillars looking for a protected place to overwinter

wigglers—mosquito larvae, their name refers to their swimming behavior

zoophagous—organisms that feed on other animals, including predators and parasites

Zoraptera—literally "pure wingless," insect order of tiny, usually wingless insects known as zorapterans

Index